TWENTY FIRST CENTURY
SCIENCE

REVISION

GCSE Science

Philippa Gardom Hulme

OXFORD
UNIVERSITY PRESS

OXFORD
UNIVERSITY PRESS

Great Clarendon Street, Oxford OX2 6DP

Oxford University Press is a department of the University of Oxford.
It furthers the University's objective of excellence in research, scholarship,
and education by publishing worldwide in

Oxford   New York

Auckland   Cape Town   Dar es Salaam   Hong Kong   Karachi
Kuala Lumpur   Madrid   Melbourne   Mexico City   Nairobi
New Delhi   Shanghai   Taipei   Toronto

With offices in

Argentina   Austria   Brazil   Chile   Czech Republic   France   Greece
Guatemala   Hungary   Italy   Japan   Poland   Portugal   Singapore
South Korea   Switzerland   Thailand   Turkey   Ukraine   Vietnam

British Library Cataloguing in Publication Data

Data available

ISBN 978-0-19-915140-0

10 9 8 7

Printed in Great Britain by Ashford Colour Press Ltd.

Author Acknowledgements
Many thanks to my husband, Barney Gardom, for his countless contributions and endless for his countless and
endless patience. Thanks, too, to Mary and Edward Hulme, for their careful checking and valuable questions.
Finally, thanks to Catherine for putting up with her Mum spending hours at the computer instead of building
brick towers and reading stories.

Acknowledgements
These resources have been developed to support teachers and students undertaking a new OCR suite of GCSE
Science specifications, Twenty First Century Science.

Many people from schools, colleges, universities, industry, and the professions have contributed to the
production of these resources. The feedback from over 75 Pilot Centres was invaluable. It led to significant
changes to the course specifications, and to the supporting resources for teaching and learning.

The University of York Science Education Group (UYSEG) and Nuffield Curriculum Centre worked in partnership
with an OCR team led by Mary Whitehouse, Elizabeth Herbert and Emily Clare to create the specifications,
which have their origins in the Beyond 2000 report (Millar & Osborne, 1998) and subsequent Key Stage 4
development work undertaken by UYSEG and the Nuffield Curriculum Centre for QCA. Bryan Milner and Michael
Reiss also contributed to this work, which is reported in: 21st Century Science GCSE Pilot Development: Final
Report (UYSEG, March 2002).

Sponsors
The development of Twenty First Century Science was made possible by generous support from:
• The Nuffield Foundation
• The Salters' Institute
• The Wellcome Trust

THE SALTERS' INSTITUTE

wellcometrust

The
Nuffield
Foundation

Contents

About this book

To parents and guardians

This book is designed to help students achieve their best in OCR's Twenty First Century Science GCSE examinations.

It includes sections on each of the areas of biology, physics, and chemistry explored by Twenty First Century Science, as well as extensive coverage of the six 'Ideas about Science' that are a vital and integral part of the course.

This book is designed to be used! Students will get the most from it if they do as many of the Workout and GCSE-style questions as possible. Many students will also find it helpful to highlight, colour, and scribble extra notes in the Fact banks.

To students

The book is in fifteen sections. There is one section for each of the biology, chemistry, and physics modules B1 to P3, as well as six sections covering 'Ideas about science'.

Each section includes:

Workout

Go through these on your own or with a friend – many of them are quite fun! Write your answers in the book. If you get stuck, look for answers in the Fact bank. The index will help you to find what you need. Check your answers at the back of the book.

Fact bank

The Fact banks summarize information from the module on no more than four pages for each module! The 'Ideas about science' Fact banks are simply conversations – read them aloud with a friend if you want to. Look in a textbook at school if there's anything else you want to find out.

Don't just read the Fact banks – highlight key points, scribble extra details in the margin or on Post-it notes, and make up ways to help you remember things. The messier your book is by the time you take your exams, the better!

You could try getting a friend – or someone at home – to test you on the Fact banks. Or make cards to test yourself, with a question on one side and the answer on the other.

GCSE-style questions

These are as much like the module tests as possible. You could work through them using the Fact bank to check things as you go, or you could do them under test conditions. The answers are in the back of the book.

In every section, content covered at Higher level only is shown like this. **H**

Skills assessment: Data analysis and Case study

Turn to pages 164–7 for a summary of essential advice on maximizing your marks in the Skills assessment tasks.

1 Use these words to finish labelling the diagram.

cell   **nucleus**   **chromosomes**   **genes**   **DNA**

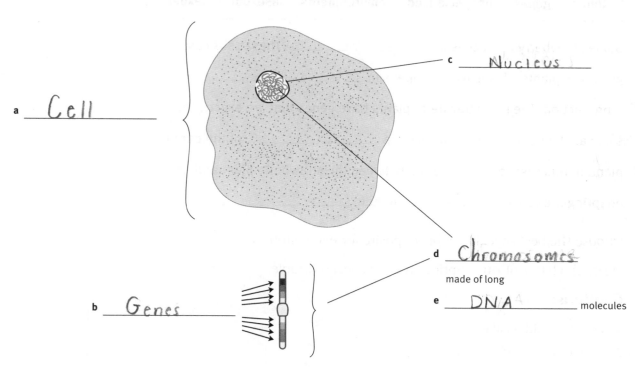

a   _Cell_

b   _Genes_

c   _Nucleus_

d   _Chromosomes_
made of long

e   _DNA_ molecules

2 Write the letter **T** next to the statements that are true.
Write the letter **F** next to the statements that are false.

a Women have two X chromosomes in each cell, except for their
sex cells. _T_

b Men have one Y chromosome in each cell. _T_

c Human egg cells contain 46 chromosomes. _F_

d Every sperm has an X chromosome. Half of all sperm also have a
Y chromosome. _F_

e If a sperm with a Y chromosome fertilizes an egg, the embryo develops
female sex organs. _____

f Your eye colour depends only on your genes, so eye colour is an
inherited characteristic. _T_

g You cannot influence characteristics that depend on both genetic and
environmental factors. _____

h Different versions of the same gene are called alleles. _T_

3   Choose words from the box to fill in the gaps.

clones	genes	unspecialized	environments	asexual	sexual

Some strawberry plant cells are _____. These cells can

grow new plants. This is what happens in _____

reproduction. The new strawberry plants have genes that are exactly the

same as their parent's genes. They are _____ of the parent

plant. In this case, all the variation between the strawberry plant and its

offspring are caused by differences in their _____.

4   Choose the best speech bubble opposite for each caption.

Write the letter of one caption in each speech bubble.

Captions:

A   I'm **h** the hairless allele

B   … and me from Dad's sperm.

C   His ring finger is hairless!

D   We're both versions of just one gene.

E   Her ring finger is hairy!

F   … and me from Dad's sperm.

G   You'll always find us in the same place on the two chromosomes of a pair …

H   … Except in sex cells – each sperm or egg cell has only one of us!

I   Clare got me from Mum's egg …

J   I'm dominant

K   So even though we've all got the same Mum and Dad, we're different from each other!

L   … and me from Dad's sperm.

M   Joe got me from Mum's egg …

N   We go round in twos.

O   Naomi got me from Mum's egg …

P   Her ring finger is hairy!

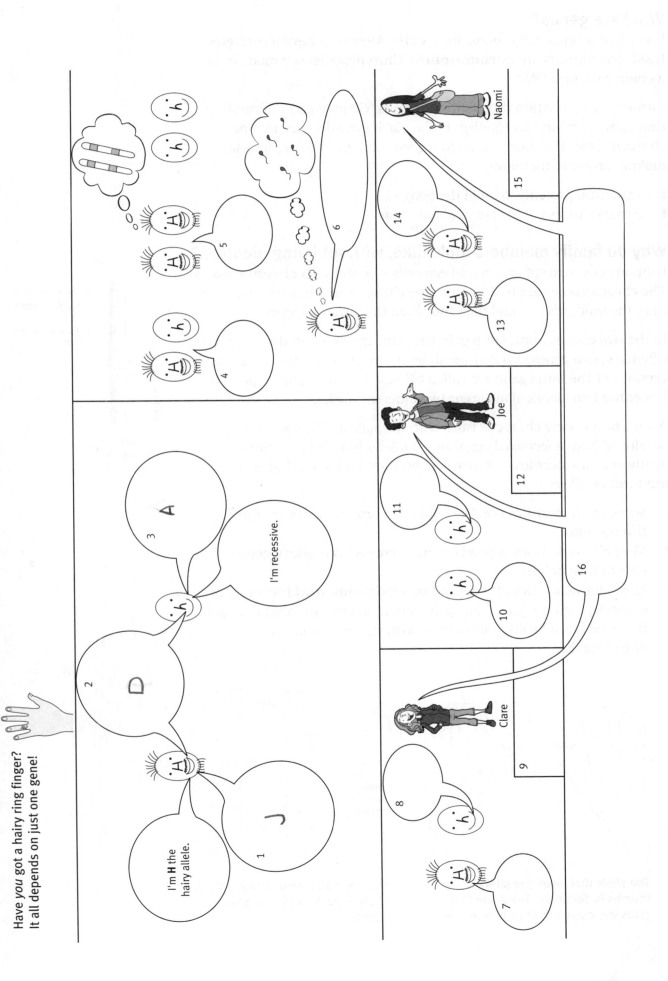

What are genes?

Every living organism is made from **cells**. Most cells have a **nucleus**. Inside the nucleus are **chromosomes**. Chromosomes are made from a chemical called DNA.

Chromosomes contain thousands of **genes**. Genes carry information that controls what an organism is like. Each gene determines one characteristic. The information in a gene is a set of instructions for making proteins, including

H
- structural proteins to build the body
- enzymes to speed up chemical reactions

Why do family members look alike, without being identical?

Human cells – except sperm and egg cells – contain **46 chromosomes**. The chromosomes are in 23 pairs. One chromosome in each pair came from the mother's egg and the other from the father's sperm.

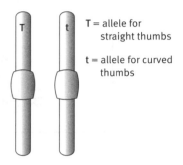

T = allele for straight thumbs

t = allele for curved thumbs

This person has straight thumbs.

In the two chromosomes of a pair, the same genes are in the same place. Often, the two genes of a pair are different from each other. Different versions of the same gene are called **alleles**. For each gene, a person has either two identical alleles or two different alleles.

A son shares some characteristics with his parents. This is because he developed from a fertilized egg that got alleles from both parents. Brothers and sisters are not identical because they have different mixtures of alleles.

- Some characteristics are determined by **one gene**, for example thumb shape.
- Most characteristics depend on instructions from **many genes**, for example height.
- Most characteristics also depend on **environmental factors**. For example, a man's genes may give him a high risk of heart disease. But if he has a healthy lifestyle he reduces his chance of getting ill.

Dominant :
An allele that produces the same phenotypic effect whether inherited with a homozygous & heterozygous

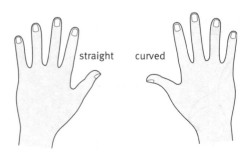

straight   curved

a father with **Tt** alleles
(straight thumbs)

sex cells	T	t
t	Tt	tt
t	Tt	tt

a mother with **tt** alleles (curved thumbs)

The allele that gives you straight thumbs is dominant. The allele that gives you curved thumbs is recessive.

There is a 50% chance that a child of these parents will have straight thumbs.

What makes humans male or female?

Humans have one pair of sex chromosomes in each body cell. Female humans have two X chromosomes (XX). Males have one X chromosome and one Y chromosome (XY).

Sex cells – sperm and eggs – contain only 23 chromosomes. The chromosomes are not paired up. Every egg has an X chromosome. Half of all sperm have an X chromosome; the other half have a Y chromosome. On the Y chromosome is a gene that is the instruction to make a male sex hormone.

At fertilization, a sperm nucleus fuses with an egg nucleus:

▸ If a Y sperm fertilizes an egg, the embryo develops male sex organs.
▸ If an X sperm fertilizes the egg, the embryo develops female sex organs.

What causes inherited diseases?

Huntington's disorder and cystic fibrosis are inherited diseases. They are caused by 'faulty' alleles of just one gene.

Huntington's disorder does not usually develop until someone is over 35. Sadly the disease is fatal.

Symptoms:

▸ loss of control over movements
▸ memory loss and mental deterioration

One faulty dominant allele – **H** – causes Huntington's disorder.

A person can inherit the disease from just one parent.

a father with **hh** alleles
(does not have
Huntington's disorder)

	sex cells	h	h
a mother with **Hh** alleles (does have Huntington's disorder)	H	Hh	Hh
	h	hh	hh

There is a 50% chance that a child of these parents will inherit Huntington's disorder.

Children with cystic fibrosis produce thick, sticky mucus. The mucus

▸ blocks the lungs and air passages, making breathing difficult
▸ prevents enzymes getting to the gut, making digestion difficult
▸ encourages bacteria to grow which causes infection

A faulty recessive allele – **f** – causes cystic fibrosis. A baby who has two faulty alleles – one from each parent – has cystic fibrosis. A baby with one faulty allele is a carrier.

Carriers do not have the disease, but can pass it on to their children.

a father with **Ff** alleles
(carrier of cystic fibrosis)

	sex cells	F	f
a mother with **Ff** alleles (carrier of cystic fibrosis)	F	FF	Ff
	f	Ff	ff

There is a 25% chance that a child of these parents will inherit cystic fibrosis. There is a 50% chance that a child will be a carrier of cystic fibrosis.

How is genetic testing used?

▶ Some adults want to know if they are **carriers of a genetic disease**. Doctors extract genes from white blood cells and test them for disease-causing alleles.

▶ Doctors can take cells from a young fetus and test them for disease-causing alleles. If, for example, a fetus has two cystic fibrosis alleles, its parents may decide on a **termination**.

H ▶ **Insurance companies** could use genetic testing to assess the risk of a person having an illness. This is not allowed in the UK.

▶ Health authorities could test a whole population for a disease-causing allele. This is **genetic screening**. It is expensive, but may be cheaper than caring for children born with the disease.

Can parents avoid having a baby with a genetic disease?

Imagine one or both members of a couple are carriers of a genetic disease. They want a baby. They may decide to use the series of techniques below:

▶ Doctors use the father's sperm to fertilize eggs outside the mother's body (*in vitro* **fertilization**). Embryos develop.

▶ Doctors test one cell from each eight-cell embryo for the disease-causing allele (**pre-implantation genetic diagnosis**).

▶ Doctors choose an embryo without the faulty allele to implant into the mother's uterus (**embryo selection**).

These techniques are not always successful.

What is gene therapy?

Scientists think some genetic diseases can be cured by gene therapy. Faulty alleles in cells will be replaced by normal alleles from a healthy person. This has worked for one disease, but not so far for cystic fibrosis.

What are stem cells? Are they useful?

Embryos contain **stem cells**. These are unspecialized cells that can develop into any type of cell. Doctors hope to use them in the future to treat some diseases.

What are clones?

Some bacteria, plants and simple animals reproduce asexually to make clones. Clones and their parents have **identical genes**. Environmental factors cause differences between clones.

Animals do not usually form clones, but there are exceptions:

▶ Identical twins are clones of each other.

▶ Scientists have made clones. They removed an egg cell nucleus. They took another nucleus from an adult body cell and transferred it to the 'empty' egg cell. They grew the embryo for a few days and then implanted it into a uterus.

1 Ellen and Hannah are identical twin girls.

a Ellen and Hannah look the same as each other.
Choose the best explanations for this.
Put ticks in the correct boxes.

They have the same combination of alleles. ☑

They inherited genes from both parents. ☐

They both developed from one egg that was
fertilized by one sperm. ☐

They both started growing from one embryo.
The cells of the embryo separated. ☐ [1]

b Ellen and Hannah look different from their mother.
Choose the best explanations for this.
Put ticks in the correct boxes.

A person's characteristics are affected by both genes
and the environment. ☐

They received alleles from both parents. ☐

The twins and their mother have different
combinations of alleles. ☐

Their cells contain 23 pairs of chromosomes. ☐ [1]

c Ellen has one pair of sex chromosomes in each body cell.
Which two chromosomes are in this pair?
Circle the correct answer.

XY   YY   XX [1]

d John and Jim are identical twins. They are 50 years old.
John is fatter than Jim. Choose the best explanation for this.
Put a tick in the correct box.

They have different combinations of alleles. ☐

They are clones of each other. ☐

They have different lifestyles. ☐

John was born an hour before Jim. ☐ [1]

Total [4]

2 Complete the following sentences about genes.
Choose from this list.

alleles   **information**   **proteins**   **characteristic**
          carbohydrates   **chromosomes**   **fats**

Living things are made from cells. Inside every cell nucleus are very

long threads called _____. These are made of

thousands of genes. Genes carry _____ that controls

how a living thing will develop. Genes are the code for making

_____. Each gene controls one _____.   [4]

3 Complete the following sentences about stem cells.
Choose from this list.

embryos   **muscles**   **specialized**   **measles**
          unspecialized   **research**

Stem cells are _____ cells. Stem cells in

_____ can develop into any type of cell. Scientists

are doing _____ about stem cells because it may be

possible to use them to treat some illnesses.   [3]

4 The allele that causes straight thumbs is dominant (**T**).
The allele that causes curved thumbs is recessive (**t**).

Sarah has straight thumbs. She has one **T** allele and one **t** allele.
Alan has curved thumbs. He has **tt** alleles.

a i What percentage of Sarah's egg cells contain the allele **T**?

_____   [1]

   ii Give the number of **t** alleles in each of Alan's body cells
(*not* the number in his sperm cells).

_____   [1]

b i Finish the diagram to show which alleles Sarah and
Alan's children may inherit.

	Sarah (mother) Tt	
sex cells	**T**	**t**
Alan (father) tt **t**		
t		

[3]

  ii Sarah and Alan have a baby boy.
What is the chance of his having a straight thumb?
Put a ring round the correct answer.

    25%   **50%**   **75%**   **100%**   [1]

c Give an example of one human feature that is affected by
several different genes.

_____   [1]

Total [7]

5 Huntington's disorder is an inherited disease.
Its symptoms usually develop after the age of 35.

a Give two symptoms of Huntington's disorder.

_____

_____ [2]

b Huntington's disorder is caused by a dominant allele of just
one gene. The table shows the alleles of this gene in the cells
of four people.

Name	Alleles
Abigail	Hh
Brenda	HH
Chris	hh
Deepa	hh

Who will develop Huntington's disorder?
Circle the correct name or names.

Abigail   **Brenda**   **Chris**   **Deepa** [1]

c Gary is 20. He had a genetic test. The test shows that he will
develop Huntington's disorder.

i Suggest one reason why Gary may not want to tell his employer
the results of the test.

_____

_____ [1]

ii Gary's wife is six weeks pregnant. Suggest one reason why the
couple may decide to test the fetus for Huntington's disorder.
Suggest one reason why they may decide not to have this test.

For the test: _____

_____

Against the test: _____

_____ [2]

Total [6]

1 Label the pie chart with the names of the gases of the Earth's atmosphere.

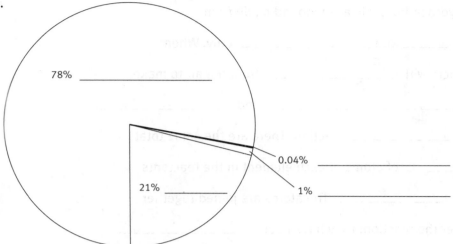

78% _____

0.04% _____

21% _____

1% _____

H **2** Write these formulas in sensible places on the drawings.

O_2 NO_2 CO_2 N_2 H_2O NO   C   CO

petrol (hydrocarbons)

My car has a catalytic converter.

A

petrol (hydrocarbons)

B

Cats didn't exist when they made this old banger!

3 a Solve the anagrams to find the names of four pollutants. Each name has two words.

b Draw one or two lines from each pollutant to show where it goes.

died oxen rioting _____	Makes surfaces dirty and causes health problems if breathed in.
lex did furious _____	Used by plants in photosynthesis.
a carbonate culprit _____	Dissolves in rainwater and sea water.
ionic adder box _____	Dissolves in rainwater and sea water and lowers the pH of rain.

4 Fill in the gaps.

Octane is a hydrocarbon. It is a compound made from

_____ and _____ only. When

it burns, it reacts with _____ from the air to make

_____ _____ and _____.

This is a _____ reaction. There are the same total

_____ of atoms of each element in the reactants

and in the _____. The atoms are joined together

differently after the reaction; they have been _____.

The properties of the _____ are different from the

properties of the reactants.

5 Fill in the empty boxes to summarize some combustion reactions of coal.

	Reactants		Product
Name	coal (with no sulfur impurities)	oxygen (from a plentiful supply of air)	
Formula	C	O_2	CO_2
Diagram			

	Reactants		Products		
Name	coal (with no sulfur impurities)	oxygen (from a limited supply of air)		carbon monoxide	particulate carbon
Formula	C		CO_2		C
Diagram					

	Reactants		Products		
Name	coal (with sulfur impurities)	oxygen (from a plentiful supply of air)			
Formula	C and S		CO_2	SO_2	
Diagram					

6 Annotate the diagram to show how to reduce atmospheric pollution from car exhaust gases.

a

b

c

d catalytic converter

e

7 Fill in the empty boxes.

Pollutant name	Pollutant formula	Where the pollutant comes from	Problems the pollutant causes	One way of reducing the amount of this pollutant added to the atmosphere
sulfur dioxide				
nitrogen oxides				
carbon dioxide				
carbon monoxide				
particulate carbon				

8 Solve the clues to fill in the arrow-word.

Across (left to right)

1 The Earth is surrounded by an . . .
12 The best way to reduce the atmospheric pollution from power stations that burn fossil fuels is to use . . . electricity.
14 Carbon dioxide is a . . . of the reaction between a hydrocarbon and oxygen.
17 Burning coal in a power station is one way of generating . . .

Across (right to left)

11 Sulfur dioxide and nitrogen dioxide react with water and oxygen to produce acid . . .
16 In chemical reactions, the atoms are . . .

Down

1 The Earth's atmosphere contains 1% . . . gas.
2 Incomplete burning of fossil fuels produces carbon . . .
3 The Earth's atmosphere contains 21% . . . gas.
4 Use low-. . . fuels to reduce atmospheric pollution from motor vehicles.

5 . . . is a mixture of hydrocarbons that fuels many cars.
6 Carbon monoxide is a pollutant that directly harms . . .
9 Much coal contains sulfur as an . . .

Up

7 The symbol for carbon is . . .
8 . . .-sulfur fuels make less sulfur dioxide when they burn than high-sulfur fuels.
10 The major products of combustion of methane (natural gas) are carbon dioxide and . . .
13 More efficient engines burn less . . . than less efficient engines.
14 Nitrogen monoxide is an example of a . . .
15 An . . . dissolves in water to make a solution with a pH less than 7.
16 In car engines, nitrogen reacts with oxygen to make nitrogen monoxide. Nitrogen and oxygen are the . . .
18 In . . . converters, carbon monoxide reacts with oxygen to make carbon dioxide.

Which chemicals are in the air?

The Earth's atmosphere is a **mixture of gases:**

78%   nitrogen

21%   oxygen

1%   argon

0.04%   carbon dioxide

and variable amounts of water vapour.

Also, human activity adds harmful chemicals – pollutants – to the air.

Earth

atmosphere

How do chemical reactions make air pollutants?

Coal is mainly **carbon**. Petrol and diesel are mainly **hydrocarbons** (compounds of hydrogen and carbon). These fossil fuels burn in power stations to generate electricity, and in vehicles like cars and trains.

When fuels burn they react with oxygen from the air. This is **combustion**. In all chemical reactions, the atoms are rearranged. The same atoms are present in both the reactants and the products; they are just joined together differently. This is the conservation of atoms.

For example, methane (natural gas) reacts with oxygen to make carbon dioxide and water.

The properties of the reactants are very different from the properties of the products. Some products of hydrocarbon combustion are pollutants.

CH_4	O_2 O_2	CO_2	H_2O H_2O
methane	oxygen	carbon dioxide	water
reactants		products	

● 1 carbon atom (C)   ● 1 carbon atom (C)

○ 4 hydrogen atoms (H)   ○ 4 hydrogen atoms (H)

◐ 4 oxygen atoms (O)   ◐ 4 oxygen atoms (O)

Which chemicals are air pollutants?

▸ sulfur dioxide, SO_2

▸ nitrogen monoxide, NO

▸ nitrogen dioxide, NO_2

▸ carbon monoxide, CO

▸ carbon dioxide, CO_2

▸ small particles of solids (particulates), for example carbon

Where do air pollutants come from?

Fossil fuels burn to make carbon dioxide. Sometimes there is not enough oxygen to convert all the carbon in the fuel to carbon dioxide. Then **incomplete combustion** occurs. This makes carbon monoxide and solid carbon.

Some fossil fuels contain sulfur impurities. When the fuel burns, the sulfur reacts with oxygen to make sulfur dioxide.

At high temperatures inside engines, nitrogen and oxygen from the air react to make nitrogen oxides.

H Nitrogen monoxide (NO) forms first. This reacts with more oxygen from the air to make nitrogen dioxide (NO_2) – in other words, nitrogen monoxide is oxidized. Together, NO and NO_2 are called NO_x.

Where do air pollutants go?

Some pollutants harm humans directly.
Others damage the environment, and so harm humans indirectly.

▶ Particulate carbon makes surfaces dirty and can cause health problems if breathed in.
▶ CO is poisonous.
▶ CO_2 dissolves in rainwater and sea water. Plants use it for photosynthesis. CO_2 in the atmosphere contributes to global warming.
▶ SO_2 and NO_2 react with water and oxygen to make acid rain. NO_2 may increase the risk of asthma attacks.

How can we improve air quality?

The only way to make less CO_2 is to burn less fossil fuel.

We can reduce air pollution from fossil fuel power stations by
▶ using less electricity
▶ removing sulfur impurities from fuels before burning them
▶ removing sulfur dioxide and particulates (carbon and ash) from the gases that power stations emit

We can reduce air pollution from vehicle exhaust gases by
▶ developing efficient engines that burn less fuel
▶ using low-sulfur fuels
▶ using catalytic converters to convert
   – nitrogen monoxide to nitrogen and oxygen
   – carbon monoxide to carbon dioxide
▶ using public transport instead of cars
▶ having legal limits on exhaust emissions

1   Natural gas is a hydrocarbon. Many people use it as a fuel
    for cooking.

   a   What is a hydrocarbon?
        Tick the one best definition.

        A hydrocarbon is a mixture of the elements hydrogen
        and carbon only.   ☐

        A hydrocarbon is a compound made from hydrogen
        and carbon only.   ☐

        A hydrocarbon is a mixture of the elements hydrogen
        and carbon.   ☐

        A hydrocarbon is a compound made from hydrogen
        and carbon.   ☐   [1]

   b   The products of the complete combustion of methane
        are carbon dioxide and water.

        i Finish the diagram to represent this reaction.

                                                                                    [1]

        ii Where does the carbon dioxide go?
            Put ticks in the correct boxes.

            It is used by animals in respiration.   ☐

            It is used by plants in photosynthesis.   ☐

            It dissolves in rainwater and sea water.   ☐

            It mixes with other gases in the atmosphere.   ☐   [2]

c Incomplete combustion of methane makes particulate (solid) carbon and carbon monoxide gas, as well as carbon dioxide.

 i Give one problem caused by particulate carbon.

 _____ [1]

 ii Match the name of each pollutant to the correct formula and picture.

Name	Formula	Picture
carbon dioxide	C	
carbon monoxide	CO	
particulate carbon	CO_2	

[3]

d What is the **one** way of producing less carbon dioxide?

 _____ [1]

Total [9]

2 a Complete the sentences to describe how oxides of nitrogen are made in car engines.

 At high temperatures inside engines, nitrogen and _____

 from the air react to make nitrogen oxides. Nitrogen _____

 forms first. This then reacts with more oxygen from the air to make

 nitrogen _____. [3]

b Give one problem that nitrogen oxides cause.

 _____ [1]

Total [4]

3 a i Sulfur dioxide is made when sulfur reacts with oxygen from the air.

Finish the diagram to represent this reaction.

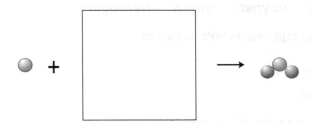

[1]

ii One source of sulfur is in the coal that power stations burn to generate electricity.

Name one other source of the sulfur that reacts to make sulfur dioxide.

_____ [1]

b The graph shows how the amount of sulfur dioxide emitted by China has changed since 1980.

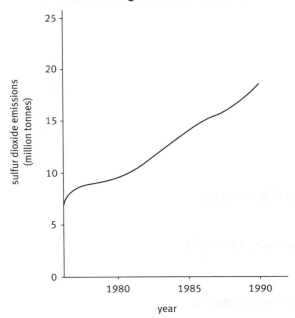

i Use the graph to complete the sentence below.

Between 1980 and 1990 the amount of sulfur dioxide emitted

by China _____. [1]

ii Suggest why the amount of sulfur dioxide emitted by China changed in this way.

_____ [1]

c i Sulfur dioxide mixes with other gases in the atmosphere.
Circle the names of the two gases in the Earth's atmosphere
that are present in the two largest amounts.

nitrogen   carbon dioxide   oxygen   argon   hydrogen   [2]

ii In the atmosphere, sulfur dioxide reacts with water to
make acid rain.
Why is acid rain a problem?
Put ticks in the correct boxes.

Acid rain damages buildings made of limestone.   ☐

Acid rain makes lakes more acidic.   ☐

Acid rain damages trees.   ☐

Acid rain increases the pH of streams.   ☐   [2]

d This graph shows how the amount of sulfur dioxide emitted
by the UK has changed since 1980.

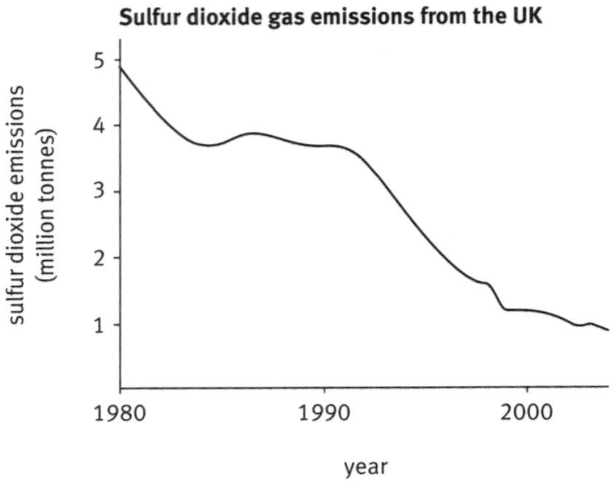

Sulfur dioxide gas emissions from the UK

What actions might have helped to achieve this change?
Put ticks in the correct boxes.

Power stations remove sulfur from coal before burning it.   ☐

British people use less electricity.   ☐

Power stations remove sulfur dioxide from waste gases.   ☐

Less electricity is being produced in power stations that
burn coal.   ☐   [2]

Total [10]

D

C

B

A

This diagram shows sedimentary rock containing fossils. Assume that this rock has never been folded.

1 Give the letter of

    a the layer that contains the youngest fossils _____

    b the layer made of the oldest rocks _____

    c the layer made of the youngest rocks _____

    d the layer in which sediments were first deposited _____

2 Write the correct numbers in the gaps. Use the numbers in the box.

3 hundred thousand	**10**	**4 thousand million**
1000	**1.4 million**	**12 700**

    a Light travels at _____ km/s.

    b The Earth's oldest rocks are _____ years old.

    c Seafloors spread by about _____ cm each year.

    d The diameter of the Earth is _____ km.

    e The diameter of the Sun is about _____ km.

    f The diameter of the biggest asteroid in our Solar System is _____ km.

3 Cross out the words that are wrong:

Distant galaxies are moving **towards/away from** us.

The graph shows that as the distance of a galaxy from the Milky Way increases, the speed at which the galaxy moves away from us **increases/decreases**. This is **Hubble's/Hutton's/Hawking's** law.

These galaxy movements mean that space is **getting smaller/ expanding**, and also provide evidence that the Universe began with a 'big bang' fourteen **hundred/thousand/million** million years ago.

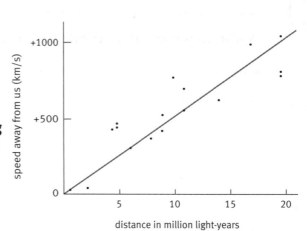

4 This question is about the possible ultimate fate of the Universe.

Label the graph below by writing one letter on each line.

  A The Universe will reach a maximum size and then stop.

  B The Universe will collapse with a 'big crunch'.

  C The Universe will expand forever.

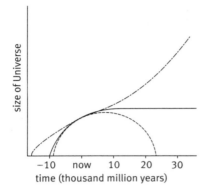

5 Finish the notes about volcanoes and earthquakes.

A **volcano** is _____

Volcanoes are usually found at _____

Signs that a volcano will erupt soon are _____

_____

Authorities can reduce the damage caused by a volcano by _____

_____

_____

An **earthquake** is _____

Earthquakes usually happen at _____

These types of movement cause earthquakes: _____

_____

_____

Authorities can reduce the damage caused by earthquakes by _____

_____

_____

What do we know about the Earth's structure, and how it changes?

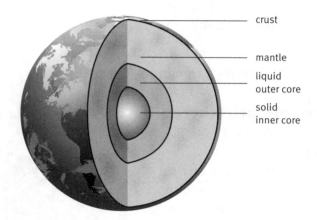

crust

mantle

liquid outer core

solid inner core

Evidence from rocks tells us about the structure of the Earth. For example, the Earth must be older than its oldest rocks. Scientists use **radioactive dating** to estimate rocks' ages.

Rock processes we see today explain past changes. For example, mountains are being made all the time – if not, **erosion** would wear down the continents to sea level.

Wegener's theory of continental drift (1912)
Wegener's theory states that today's continents were once joined together as one huge continent. They have been slowly moving for millions of years.

See page 134 for more details about Wegener's theory.

250 million years ago

Theory of seafloor spreading (1960s)
Geologists detected **oceanic ridges**, which are lines of mountains under the sea. They noticed a symmetrical stripe pattern in rock magnetism on each side of these ridges. They devised this explanation:

New ocean floor is made at oceanic ridges, so oceans spread by about 10 cm a year.

The new ocean floor is made like this:

Hot mantle rises beneath the ridge. It melts to make magma. Magma erupts at the middle of the ridge. It cools to make new rock. The new rock is magnetized in the direction of the Earth's field at the time.

The theory of plate tectonics (1967 onwards)

The outer layer of the Earth is made of about 12 huge pieces
of rock, called **tectonic plates**. They move slowly all the time.
Earthquakes, volcanoes, and mountain building usually
happen where tectonic plates meet.

▶ **Plates move apart** at oceanic ridges. Molten rock (magma) rises up
  between the plates.

▶ In the Himalayas, **plates move towards each other**.
  They collide. Huge pressure makes rocks fold over on
  top of each other to **build mountains**.

▶ Most **volcanoes** are at plate boundaries where the crust
  is stretching or being compressed. Magma erupts out of a
  hole in the Earth's surface. Geologists monitor volcanoes
  carefully. They look for changes in the gases emitted and
  the swelling of a volcano's sides. If a volcano is likely to
  erupt the government may evacuate the area.

▶ Most **earthquakes** happen at rock breaks, called faults.
  The blocks of rock on each side of the fault move.
  Pressure builds up until the rocks snap.

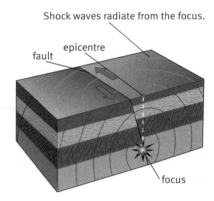

Shock waves radiate from the focus.

epicentre

fault

focus

In some places at risk from earthquakes, buildings must be
built to withstand earthquake damage. Scientists cannot
predict when earthquakes will strike.

▶ The movement of tectonic plates contributes to the **rock cycle**.

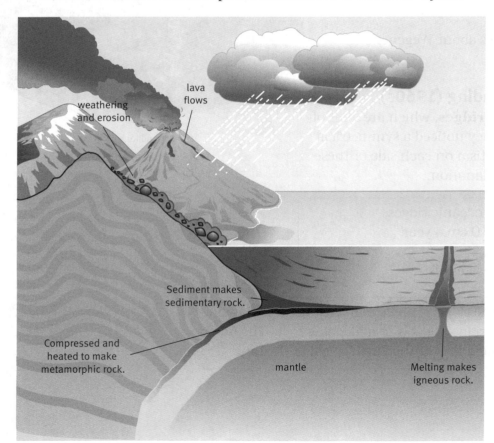

weathering
and erosion

lava
flows

Sediment makes
sedimentary rock.

Compressed and
heated to make
metamorphic rock.

mantle

Melting makes
igneous rock.

What do we know about our Solar System?

Our **Solar System** was formed from clouds of gases and dust in space.

The **Sun** is a star at the centre of the Solar System. It will probably shine for another 5000 million years. A **star** is a ball of hot gases, mainly hydrogen. In stars like the Sun, hydrogen nuclei join together (fuse) to make helium nuclei. This is the source of stars' energy. Stars change over time – they have a **life cycle**.

Eight **planets**, including Earth, orbit the Sun. Pluto used to be regarded as the ninth planet in our Solar System, but in August 2006 the scientists of the International Astronomical Union voted to reclassify Pluto as a 'dwarf planet'. Some planets have natural satellites (**moons**) that orbit them. **Comets** are lumps of rock held together by ice and frozen gases.

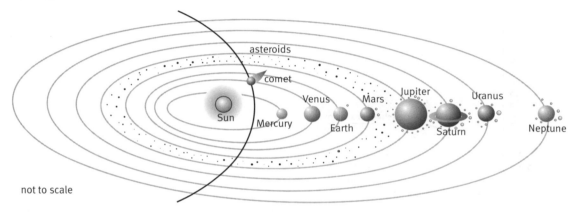

Asteroids are lumps of rock that are much smaller than planets. Most asteroids orbit the Sun between Mars and Jupiter, but a few cross the Earth's path. There is a tiny risk that one of these asteroids will collide with Earth. If this happens, many people will die. An asteroid collision may have led to the extinction of dinosaurs 65 million years ago.

Type of body	Diameter	Age
planet	Mercury = 4880 km (smallest) Saturn = 120 000 km (biggest) Earth = 12 700 km	The Solar System was formed about 5000 million years ago.
moon	Earth's Moon = 3500 km Moons are smaller than the planets they orbit.	The Earth is older than its oldest rocks, which are 4000 million years old.
asteroid	up to 1000 km; most are much smaller.	
comet	a few km	
Sun	1.4 million km	5000 million years
Universe	many millions of times greater than the diameter of the Solar System	14 000 million years

What do we know about stars, galaxies, and the Universe?

Our Solar System is part of the Milky Way galaxy. **Galaxies** contain thousands of millions of stars. The **Universe** is made of thousands of millions of galaxies.

Other galaxies are moving away from us, because space is expanding. Hubble discovered that galaxies that are further away from us move faster than those that are closer to us. This is evidence that the Universe started with a '**big bang**'.

We do not know what will happen to the Universe; scientists disagree about how to interpret evidence about its final fate. Maybe the Universe will continue to expand. Or perhaps the force of gravity will attract galaxies towards each other again and the Universe will end with a 'big crunch'.

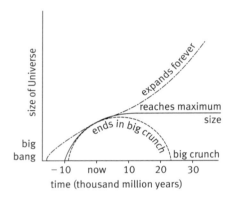

Is there life elsewhere in the Universe?

Scientists have detected planets around some stars. Life may exist on other planets in the Universe, but there is so far no evidence for this.

How do scientists find out about distant stars and galaxies?

Scientists can learn about other stars and galaxies only by studying the radiation they emit. They measure the distance to stars by looking at their relative **brightness** or by **parallax**. It is difficult to make accurate observations, so scientists do not know exact distances between objects in space.

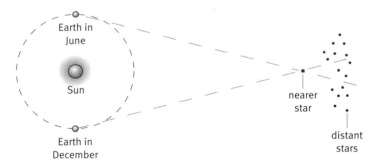

Light travels at 300 000 km/s. So when scientists observe distant objects they see what the object looked like when the light left the object – not what it looks like now. Scientists measure distances in space in light-years. One light-year is the distance light travels in one year.

Light pollution near cities makes it difficult to see stars, so scientists set up telescopes in areas of darkness away from cities.

The Earth moves from one side of the Sun to the other. Nearby stars seem to move compared to the background of distant stars. The nearer a star is to Earth, the more it seems to move. This is **parallax**.

1 a Write the following in order of age:

Earth   **Sun**   **Universe**

oldest: _____

_____

youngest: _____   [1]

b Draw straight lines to match each **type of body** to its **description**.

Type of body
moon
comet
asteroid
star

Description
a body that looks like a small, rocky planet
a natural satellite
a ball of hot gases
a lump of rock held together by ice

[3]

c 2000 years ago, a Chinese astronomer, Gan Dej, saw Ganymede, one of Jupiter's moons. He did not have a telescope or other optical instrument.

Suggest why it was easier to see Ganymede without a telescope 2000 years ago than it is now.

_____   [1]

d Use the information in the box and the table to answer questions **i** and **ii**.

In August 2006, the scientists of the International Astronomical Union defined a planet as a body that orbits a star, is big enough to have a spherical shape, and has no other objects in its orbit. They also identified four 'dwarf planets', including Pluto.

Name of dwarf planet	Approximate diameter (km)	Object it orbits	Other information
UB313 (Xena)	3000	Sun	It was made from gases and dust when the Solar System began.
Pluto	2400	Sun	Its orbit overlaps Neptune's orbit.
Ceres	1200	Sun	It is the biggest object in the asteroid belt.

i Give one way in which UB313 *does* fit the definition of a planet.

_____

ii Give one way in which Pluto *does* not fit the definition of a planet.

_____   [2]

Total [7]   31

2 a On the diagram of the Earth, label the **crust**, **mantle**, and **core**.

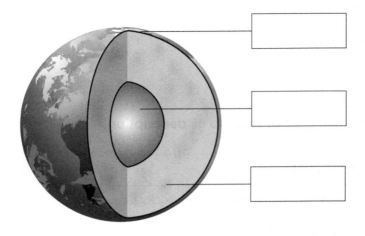

[2]

b Draw straight lines to match each **explanation** with **evidence** that supports it.

Explanation	Evidence
1 South America and Africa were once part of one big continent.	**A** Scientists have found many craters.
2 Mountains are being formed all the time.	**B** Radioactive dating of rocks.
3 The Earth is older than four thousand million years.	**C** Scientists have found the same fossils on both sides of the Atlantic Ocean.
4 Asteroids have collided with Earth.	**D** Rocks are continually eroded but the continents are not all at sea level.

[3]

Total [5]

3 In the year 964, a Persian astronomer called Abd al-Rahman al-Sufi described Andromeda galaxy. He noticed that it gave out light.

a Give the name of the galaxy of which our Solar System is a part.

_____

[1]

b How many stars are there in a typical galaxy?
Put a ring round the correct answer.

thousands of millions   **about a million**
millions of millions   **about a thousand**

[1]

c   Why does Andromeda galaxy give out light?
    Put a tick in the box next to the best answer.

    There are many planets in Andromeda. These give out light. ☐

    There are many moons in Andromeda. These give out light. ☐

    Andromeda reflects light from our Sun. ☐

    The stars of Andromeda emit light. ☐   [1]

d   Abd al-Rahman al-Sufi recorded the positions and colours
    of some of Andromeda's stars.

    i   Why are the stars different colours?

    _____

    _____

    _____   [1]

    ii   Today, scientists still cannot be certain of the distances
         of stars and galaxies from Earth.
         Put ticks in the boxes next to the **best two** reasons
         for this uncertainty.

         Everything we know about stars and galaxies comes
         only from the radiation they emit.   ☐

         Scientists use a star's relative brightness to measure
         its distance from Earth. But relative brightness also
         depends on how much dust there is between the
         star and the Earth.   ☐

         Scientists use a star's relative brightness to measure
         its distance from Earth. But relative brightness also
         depends on what stage in its life cycle a star is at.   ☐

         It is impossible to measure distances when it is cloudy.   ☐   [2]

    iii The distance from Andromeda to Earth is approximately
         2 million light-years.

         What is a light-year?

    _____

    _____   [1]

                                                                    Total [7]

33

H **4** **a** At what rate do seafloors spread?
Put a ring round the best answer.

1 cm/year   **10 cm/year**   **1 m/year**   **1 km/year**   [1]

b Put the letters of the statements in the box in a sensible order to create a paragraph that describes and explains an observation.

_____ because _____ and _____.

A the Earth's magnetic field reverses regularly
B there is a symmetrical pattern in the magnetism recorded on either side of oceanic ridges
C magma erupts, cools, and solidifies at the middle of oceanic ridges

[1]

c Draw straight lines to match each **event** with **how it happens**.

Event	How it happens
1 Mountain building	**A** Magma comes out of a hole in the Earth's surface at a place where the crust is being made or destroyed.
2 Earthquakes	**B** Tectonic plates move towards each other and collide. Massive pressure makes the rocks fold over on top of each other.
3 Volcanoes	**C** Blocks of rock on either side of a fault move. Pressure builds up and the rocks eventually snap.

[2]

Total [4]

1 **a** Solve these anagrams:

we sat   **sink**   **stare**   **a scotch maid**

b Use your answers to annotate the picture to show some natural barriers to harmful microorganisms entering the body.

2 Write each phrase from the box in a sensible place on the flow diagram.

damage cells	**disease symptoms**
reproduce rapidly	**make toxins**

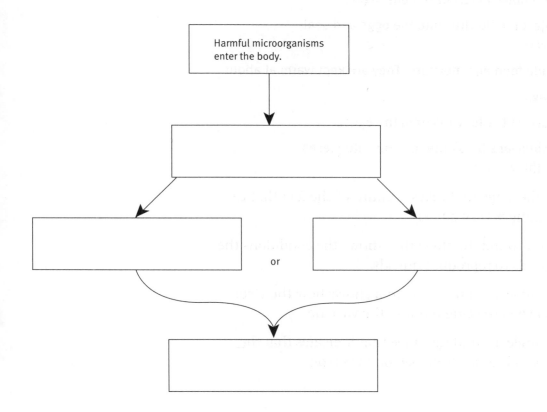

Harmful microorganisms enter the body.

or

3 The stages below describe how a vaccine works.

A The vaccine is made from dead or inactive parts of disease-causing microorganisms.

B White blood cells digest the clump.

C The vaccine is injected.

D When an active form of the same microorganism enters the blood, white blood cells make the same antibodies again. The microorganism is quickly destroyed.

E White blood cells make antibodies that stick to the microorganisms.

F The antibodies make the microorganisms clump together.

The stages are in the wrong order.

Write a letter in each empty box to show the correct order.

A					

4 Here are the stages in making an influenza (flu) vaccine.

▸ Experts meet in April to decide which strain of wild flu virus is likely to attack next winter.

▸ In labs, scientists make a special 'hybrid' flu virus.

▸ This flu virus is delivered around the world.

▸ Technicians drill holes in fertilized hens' eggs.

▸ Technicians inject the flu virus into the eggs and seal the hole with wax.

▸ The eggs provide food and moisture. They are kept warm at about 37 °C for 10 days.

▸ Technicians harvest the flu virus from the eggs.

▸ In October, technicians break the flu virus into pieces and put it into the vaccine.

    a Underline the stage that takes account of the fact that the flu virus changes very quickly.

    b Draw a box around the stage that shows the conditions the flu virus needs to reproduce quickly.

    c Draw a cloud around the stage that shows how the virus is made safe before being put into the vaccine.

    d Draw a triangle around the stage that indicates that the flu virus spreads easily from person to person.

5 Write the letter **T** next to the statements that are true.
Write the letter **F** next to the statements that are false.

a Antibiotics kill fungi and bacteria. _____

b New drugs are tested for effectiveness on human cells that were grown in the lab. Then they are tested for safety on healthy human volunteers. _____

c New drugs are tested for safety and effectiveness on people who are ill. _____

d When drugs are tested on people who are ill, one group of patients takes the new drug. Another group of patients are controls. _____

e In 'double-blind' human drug trials, doctors know who is taking the new drug and who is in the control group. _____

6 a Fill in the empty boxes about the parts of the circulation system.

Part of circulation system	What does it do?	What is it made from?
heart	It pumps blood around your body.	
artery		
vein		Veins are tubes. They have thin walls made of muscle and elastic fibres.

b i What is a heart attack?

_____

_____

_____

ii Your genes and your lifestyle both influence whether you might have a heart attack. Give three things a person can do to reduce their risk of a heart attack.

_____

_____

_____

How do our bodies resist infection?

Harmful **microorganisms** reproduce quickly inside the body, where it is warm and they have enough water and food. These conditions are ideal for microorganisms. They cause disease symptoms if they **damage cells** or **make poisons** that damage cells.

The body has barriers to microorganisms, including

- the skin
- chemicals in tears, sweat, and stomach acid

If harmful microorganisms get into the body, the **immune system** defends against the invaders. It tries to destroy them before they cause illness:

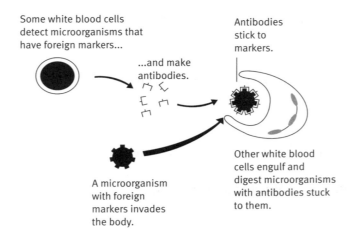

Some white blood cells detect microorganisms that have foreign markers...

...and make antibodies.

Antibodies stick to markers.

A microorganism with foreign markers invades the body.

Other white blood cells engulf and digest microorganisms with antibodies stuck to them.

Your body makes a different antibody to recognize every type of microorganism that enters it. Some of the white blood cells that make each antibody stay in your blood. So if a microorganism invades your body a second time, you quickly make the correct antibody. Your body is protected against that microorganism.

How do vaccines work?

Vaccines **prevent you getting diseases**. A vaccine contains dead or inactive parts of the disease-causing microorganism. After a vaccine is injected into your body, your white blood cells make antibodies against the microorganism. If an active – and dangerous – form of the same microorganism enters the blood in future, you make the same antibodies again. The microorganism is quickly destroyed. You are **immune** to this microorganism.

It is difficult to decide whether to have certain vaccinations. People must balance the risks of the disease against the risks of the vaccine's side effects.

H For society as a whole, vaccination is the best choice. A high percentage of the population must be vaccinated to prevent epidemics of infectious diseases.

Why is it difficult to make vaccines against some diseases?

▶ The flu virus changes quickly. There are many different strains of the disease. So new vaccines are needed every year.
▶ The HIV virus that causes AIDS changes (mutates) quickly inside the body. The virus also damages the immune system. There is no effective HIV/AIDS vaccine.

What are antibiotics, and how do they become less effective?

You cannot be immunized against every dangerous microorganism. Some microorganisms make you ill before your immune system destroys them. If the invading microorganisms are bacteria or fungi, doctors can often use antibiotics to kill them.

There are problems with antibiotics:

▶ Over time, some bacteria and fungi become resistant to antibiotics.
▶ You must take antibiotics only when necessary and finish all the tablets, even if you feel better.

Random changes (mutations) in bacteria or fungi genes make new varieties that are less affected by an antibiotic. Some of the new varieties survive a course of antibiotics.

How are new drugs developed and tested?

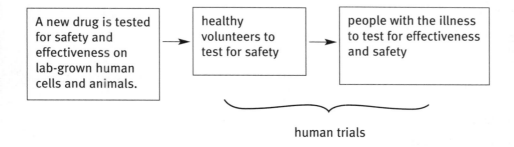

human trials

In most human trials on ill people, one group of patients takes the new drug. Another group of patients are controls. The controls take either the existing treatment for the illness, or a placebo.

A **placebo** looks like the new treatment, but has no drugs in it. Placebos are not often used in human trials because people who take them miss out on the benefits of both new and existing treatments.

Human trials are 'blind' or 'double-blind'.

▶ In **double-blind** trials, neither patients nor doctors know who is in which group.
▶ In **blind trials**, doctors know who is in which group, but patients do not.

Why does the heart need its own blood supply?

Your heart pumps blood around the body, so heart muscle cells need a continuous supply of energy. This energy comes from respiration. Respiration is a chemical reaction in cells. The reaction uses glucose and oxygen. Energy is released. Blood brings glucose and oxygen to the heart so that the heart needs its own blood supply. Coronary arteries supply blood to the heart.

How does blood travel around the body?

Blood travels around your body through **arteries** and **veins**. Most arteries carry blood away from your heart. Most veins carry blood

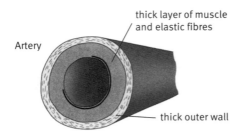

towards your heart. Blood vessels are well adapted to their functions:

Artery:

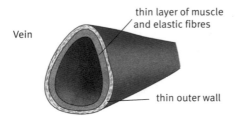

- thick, elastic muscular wall
- thin space for blood to flow through

Vein:

- thin outer wall
- wider space for blood to flow through

What is a heart attack? What factors increase the risk of heart disease?

Coronary arteries carry oxygenated blood to the heart. Fat can build up on the artery walls. A blood clot may form on this fat. This may block the artery. The blockage stops oxygen getting to the heart muscle. Heart cells die, and the heart is permanently damaged. This is a **heart attack**.

Poor diet, smoking, excess alcohol, and stress increase the risk of heart disease. Taking regular exercise reduces the risk of heart disease.

Heart disease is more common in the UK than in less industrialized countries.

1 a Catherine was coughing a lot. The doctor said she had an infection in her windpipe. He did not prescribe antibiotics.

Why might the doctor have decided not to prescribe antibiotics?

Put a tick next to the **one best** answer.

The cough was caused by a virus. ☐

She had not had the cough for very long. ☐

The cough was caused by a fungus. ☐

The cough was bad only at night. ☐ [1]

b A few days later, Catherine had a painful ear.
The doctor examined her and prescribed antibiotics.

i How did the antibiotics work?

Put a tick next to the **one best** answer.

They killed the virus that caused the painful ear. ☐

They immunized Catherine against the microorganism that caused the painful ear. ☐

They increased the resistance of the microorganism that caused the painful ear. ☐

They killed the bacteria that caused the painful ear. ☐ [1]

ii On the label, it says that you must take all the antibiotics, even if you feel better.
Explain why.

_____

_____

_____

_____ [3]

Total [5]

2 **a** Which of these lifestyle factors increase a person's risk of having a heart attack?

Put a tick in **each of the correct boxes**.

a diet low in cholesterol ☐

smoking cigarettes ☐

drinking too much alcohol ☐

not taking regular exercise ☐ [2]

b Why do heart muscle cells need their own blood supply?

Put a tick in **each of the correct boxes**.

The heart needs a continuous supply of energy to pump blood around the body. ☐

Heart muscle cells need a constant supply of oxygen and carbon dioxide for respiration. ☐

Blood brings a constant supply of glucose and oxygen to the heart. ☐

Each heart muscle cell has a nucleus that contains genetic information. ☐ [2]

c The stages of one type of heart attack are given below.

A The heart muscle is starved of oxygen.

B Fat sticks to the wall of the coronary arteries.

C Part of the heart muscle is permanently damaged.

D Blood cannot get to part of the heart muscle.

E A blood clot forms on the fat.

The stages are in the wrong order.
Write a letter in each empty box to show the correct order.

B				

[3]

Total [7]

3 Lauren has food poisoning. She has diarrhoea and vomits frequently.

She became ill after she ate a raw egg that contained
Salmonella bacteria.

a Name one natural barrier in the body that could have prevented
Salmonella bacteria entering Lauren's intestines.

_____   [1]

b The graph shows the changes in the number of *Salmonella* bacteria
in Lauren's stomach.

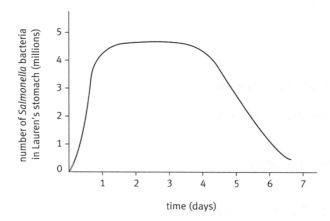

i Name the process that causes the number of bacteria to increase
during the first few hours.

_____   [1]

ii Use the graph to complete the sentence below.

Lauren will probably begin to feel better _____ days after she ate
the raw egg.   [1]

c Lauren's body tries to get rid of *Salmonella* bacteria in two ways:

> ▶ Vomiting and diarrhoea remove some of the bacteria
>   from the intestines.

> ▶ Certain blood cells can destroy the bacteria.

i Name the type of blood cells that can destroy *Salmonella* bacteria.

_____

[1]

ii Suggest why Lauren's doctor advised her **not** to take
anti-diarrhoea tablets.

_____

_____

_____

[1]

d Lauren's doctor **did not** treat her *Salmonella* with antibiotics.

Some farmers **do** give their chickens antibiotics when they
are infected by *Salmonella*.

Complete the sentences below to explain a problem
this caused. Choose words from this list.

resistant   mutated   killed   bacteria   viruses

Some *Salmonella* bacteria became _____ to

the antibiotic that farmers gave chickens. This antibiotic no

longer _____ the bacteria. So scientists searched

for new antibiotics. But eventually _____ became

resistant to them, too.

[3]

Total [8]

4   The graph shows the percentage of British 2-year-olds who had received the MMR vaccine from 1989 to 2002. The MMR vaccine prevents people getting measles, mumps, and rubella.

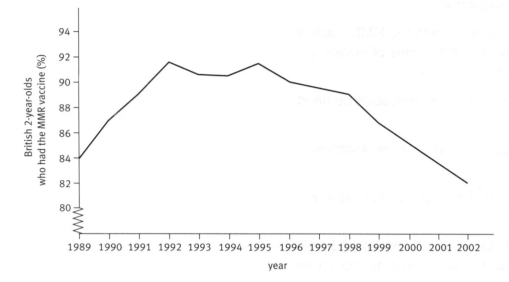

a   Here are some people's opinions about the triple MMR vaccine.

A   I'm worried about the vaccine's possible serious side effects on my child.

B   I'm a doctor. No vaccine is completely safe. The side effects of the MMR vaccine are a possible risk, but the dangers of measles, mumps, and rubella are worse.

C   Measles is a nasty disease. I don't want to risk my child getting it.

D   The more children who have the MMR vaccine, the better. Then everyone is protected from measles, mumps, and rubella.

Use the graph and the opinions to complete the sentences below.

Between 1989 and 1992 the percentage of children who had

the MMR vaccine _____. One opinion that may explain

this trend is opinion _____.

Between 1992 and 2001 the percentage of children who had

the MMR vaccine _____. One opinion that

may explain this trend is opinion _____.   [2]

b Matthew had the MMR vaccine when he was one.
Two years later, the measles virus got into his body.

Matthew did not get measles.
The stages below explain why.

A A nurse injects Matthew with the MMR vaccine.
The vaccine contains safe forms of measles,
mumps, and rubella viruses.

B Matthew's body makes the antibodies he needs
very quickly.

C The natural measles virus gets into Matthew's
bloodstream.

D The virus is destroyed before it has time to make
Matthew feel ill.

E Matthew's white blood cells make antibodies to
recognize measles, mumps, and rubella viruses.

The stages are in the wrong order.
Write a letter in each empty box to show the correct order.

A				

[3]

c Give one reason why a high percentage of the population must
be vaccinated against mumps to prevent a mumps epidemic.

_____

_____

_____ [1]

Total [6]

1 Look at the drawing of a child's tricycle.
Fill in the table.

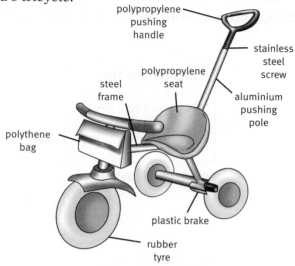

polypropylene
pushing
handle

stainless
steel
screw

polypropylene
seat

steel
frame

aluminium
pushing
pole

polythene
bag

plastic brake

rubber
tyre

Part of tricycle	Properties this part of the tricycle must have	Material
tyres		
brake		
frame		
seat		
handle to push tricycle		
pushing pole		
screws that join pushing pole to handle		
bag		

2 Use the words and phrases in the box to finish the flow diagram.

recycle	cradle	reuse	grave
landfill	use	using the product	

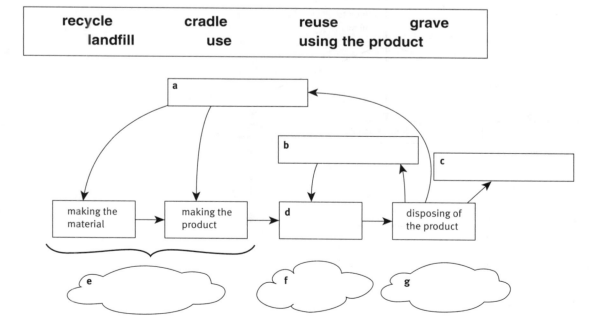

a

b

c

making the
material

making the
product

d

disposing of
the product

e

f

g

3 The products below can all be made from polymers.
Draw lines to match each product with the issues in using and
disposing of it.

Product
packaging – polythene bags
toys
water bottles

Using the product
often used just once
often used just once
used many times

Disposing of the product
can recycle, but must separate from other waste
can reuse
easy to reuse – give to someone else or a charity shop

4 Cross out the words that are wrong.

The properties of solid materials depend on how their atoms are
arranged and held together. A material has a high melting point if
the forces between one atom or molecule and its neighbours are
weak/strong. The amount of energy needed for the atoms or
molecules to break out of the solid structure is then **high/low**.

5 Use the words and phrases in the box to finish the flow diagram.

fuels   methane   petroleum jelly raw materials for chemical synthesis

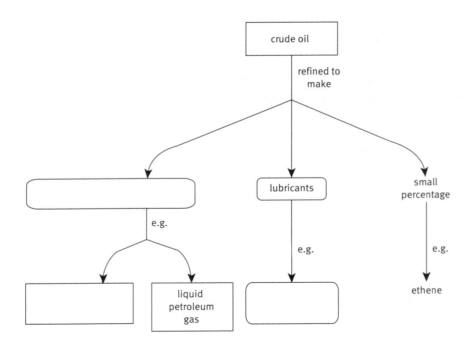

crude oil

refined to make

lubricants

small percentage

e.g.

liquid petroleum gas

e.g.

e.g.

ethene

6 Draw lines to match each modification to **one** diagram and **one or more** changes in properties.

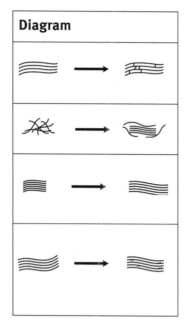

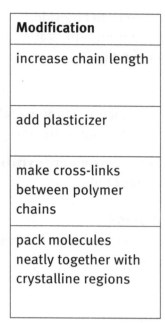

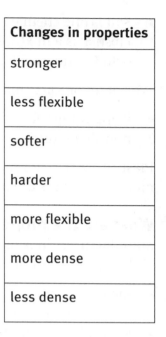

Diagram	Modification	Changes in properties
	increase chain length	stronger
		less flexible
	add plasticizer	softer
	make cross-links between polymer chains	harder
		more flexible
	pack molecules neatly together with crystalline regions	more dense
		less dense

7 Write a C in the boxes next to the equations for combustion reactions.
Write a P in the boxes next to the equations for polymerization reactions.

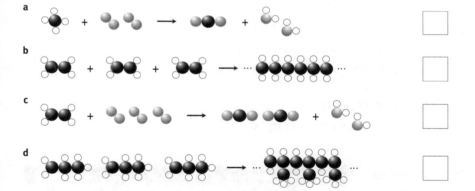

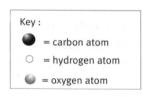

Key :
● = carbon atom
○ = hydrogen atom
● = oxygen atom

8 For each equation, draw more oxygen, carbon dioxide, or water molecules in the box so that there are the same number of atoms of each element in the products and reactants.

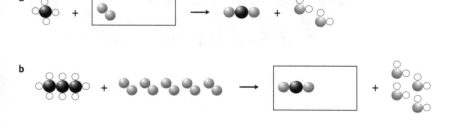

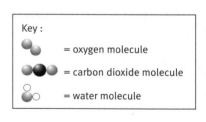

Key :
= oxygen molecule
= carbon dioxide molecule
= water molecule

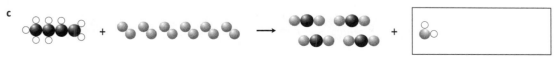

What are materials made from?

Every material is a chemical or mixture of chemicals. For example

- Salt is one chemical, the compound sodium chloride.
- Lipstick is a mixture of chemicals, including the compounds octadecanoic acid, titanium dioxide, and mica.

We obtain or make materials from

- living things, for example cotton, wool, leather, and wood
- non-living things, for example limestone, diamond, and gold

Plastics are used for many products. Some plastics are **synthetic** materials. Others are modified natural materials.

What are the properties of materials?

Manufacturers look at materials' properties to choose the best material to make a product from. Some properties are

- melting point
- strength in tension (pulling)
- strength in compression (squashing)
- stiffness
- hardness
- density

The **effectiveness** and **durability** of a product depend on the properties of the material it is made from. A product that is durable lasts for a long time before breaking, rotting, or becoming useless in some other way.

Why do materials have particular properties?

A material's properties depend on how its particles are arranged and held together. For example

- Very strong forces hold together the atoms in solid iron. An atom needs lots of energy to break out of this structure. So iron must get very hot before it becomes a liquid – it has a high melting point.
- The molecules in rubber can slide over each other. So rubber bends easily.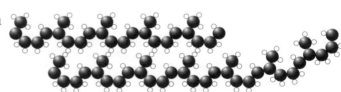

What is crude oil?

Crude oil is a mixture of **hydrocarbon** molecules of different lengths.

Hydrocarbons have molecules made from hydrogen and carbon only, for example propane and octane:

The petrochemical industry refines, or separates, the components of crude oil to make:

▶ fuels, for example petrol, diesel, and kerosene
▶ lubricants, for example Vaseline
▶ raw materials for chemical synthesis, including those to make polymers

propane

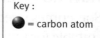

octane

Key :
● = carbon atom
○ = hydrogen atom

What are polymers?

Polymers are very long molecules containing thousands of atoms. They are made when many small molecules join together. This type of chemical reaction is called **polymerization**. There are many different polymers, all made from different starting materials.

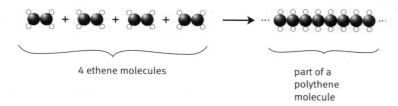

4 ethene molecules

part of a polythene molecule

In polymerization reactions – as in all chemical reactions – there are the same numbers of atoms of each element in both the reactants and the products. The atoms are **rearranged**.

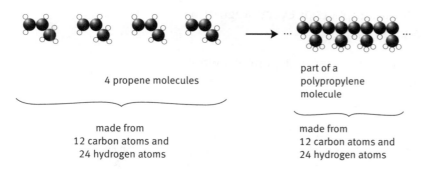

4 propene molecules

part of a polypropylene molecule

made from 12 carbon atoms and 24 hydrogen atoms

made from 12 carbon atoms and 24 hydrogen atoms

There are many polymers, each with their own unique properties, so manufacturers use polymers to make many products. Polymers replace older materials in many modern products. For example, some ropes are now made with polypropylene or nylon instead of sisal.

How do scientists change polymers' properties?

Sometimes scientists want to change a polymer's properties so that it is more suitable for making a certain product. The scientists then need to know

▶ what the polymer molecules are made from
▶ how the polymer molecules are arranged
▶ why this arrangement gives certain properties
▶ how to change the molecule arrangement to change these properties

The table shows methods of changing polymers' properties.

Method	How properties change	Diagram
making chains longer	• stronger	
adding cross-links	• harder • stronger • less flexible	
adding plasticizers	• softer • more flexible	
⊞ increasing crystallinity by lining up polymer molecules	• stronger • denser	

What is a life cycle assessment?

Manufacturers and others use a life cycle assessment (LCA) to analyse the energy use and the environmental impacts of every stage of a product's life. The diagram shows an LCA.

⊞ The outcomes of an LCA for a particular material depend on what product is made from the material.

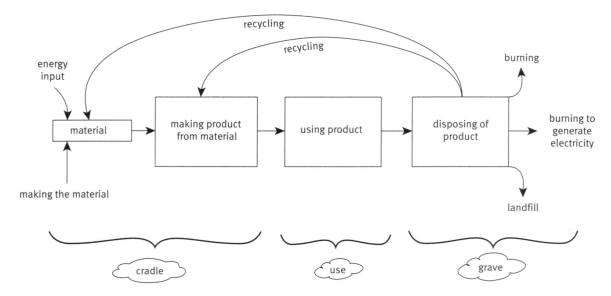

1 The table shows the properties of some synthetic polymers.

Letter	Name of material	Properties
A	poly(2-hydroxyethylmethacrylate) (PHEMA)	transparent; absorbs water to become flexible and jelly-like
B	acrylic	easily moulded into shape
C	polyethenol (PVA)	flexible, soluble in water
D	silicone rubber	insoluble in water; very durable

a Study the properties of the materials in the table.
Choose the best material to make each of the following items. Write the letter of one material next to each item.

artificial heart valves ☐

hospital laundry bags that dissolve in a washing machine, allowing dirty sheets to be washed without needing to be handled ☐

part of a composite used to make fillings for front teeth ☐

contact lenses ☐ [4]

b Disposable nappies are made from several materials, including

cellulose   **polypropylene**   **polythene**

i From the list above, write the name of one material that is obtained from a living thing.

_____ [1]

ii Draw rings round the properties that the outermost layer of a disposable nappy must have.

non-toxic   **hard**   **flexible**   **high strength in tension**   **stiff** [2]

iii Polythene is made when small molecules join together to make very long molecules. Give the name of this process.

_____ [1]

Total [8]

2 The table shows the properties of two materials:
polypropylene and sisal.

Property	Polypropylene	Sisal
durability	does not rot	rots
colour	can be pigmented any colour	can be dyed any colour
relative strength in tension	1.4	0.8
flexibility	very flexible	very flexible

a Ropes for life buoys near rivers and lakes used to be
made from sisal.
Now these ropes are made from polypropylene.
Use the table to suggest two reasons why polypropylene
is now preferred for making these ropes.

_____

_____ [2]

b The life cycle assessment (LCA) for making ropes from
polypropylene is different from the LCA for making ropes from sisal.
In what ways are the LCAs different?

Tick each correct answer.

the environmental impact of making rope from each material ☐

the environmental impact of using rope made
from each material ☐

the environmental impact of disposing of rope
made from each material ☐ [2]

c There are several methods of disposing of polypropylene ropes.

Draw straight lines to match each method to its definition.

Method	Definition
reuse	melt down and remould to make something else
recycle	burn in an incinerator that generates electricity
recover energy	put in a hole in the ground
landfill	use the rope for another purpose, for example training lifeguards

[2]

Total [6]

3 Liquid propane gas is a fuel. It is obtained from crude oil.

a Liquid propane gas burns in a bus engine.

The diagram shows how the atoms are rearranged.
The diagram is not finished.

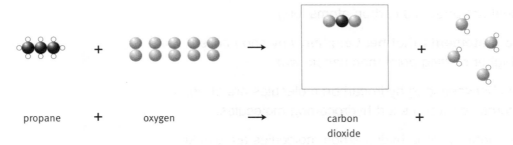

| propane | + | oxygen | → | carbon dioxide | + | |

i Complete the following sentences.

The diagram shows that when propane burns, one molecule

of propane reacts with _____ molecules of

oxygen. The products of the reaction are carbon dioxide

and _____.   [2]

ii The diagram shows that there are 3 carbon atoms in
the reactants.

What is the total number of carbon atoms in the products?

_____   [1]

iii Finish the diagram by drawing more carbon dioxide
molecules in the box so that the total number of
carbon dioxide molecules is correct.   [1]

b i Crude oil is a mixture of hydrocarbons.

What is a hydrocarbon?

_____

_____   [1]

ii Draw rings around two types of materials that are
obtained from crude oil.

lubricants   paper and card

raw materials for chemical synthesis   glass   [2]

Total [7]

4 a The chemicals used to make candle wax and shoe polish are obtained from crude oil.

▮ Candle wax is made from hydrocarbon chains that are about 30 carbon atoms long.

▮ The waxy ingredient of shoe polish is made from hydrocarbon chains that are about 70 carbon atoms long.

Tick the **two statements that best explain** why shoe polish wax has a higher melting point than candle wax.

The forces between long hydrocarbon molecules are stronger than the forces between short hydrocarbon molecules. ☐

The forces between long hydrocarbon molecules are weaker than the forces between short hydrocarbon molecules. ☐

The stronger the forces between molecules, the harder it is to separate them. ☐

The stronger the forces between molecules, the more energy is needed to separate them. ☐   [2]

b i PVC is a polymer. It is used to make these items:

▮ window frames

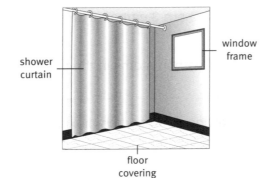

▮ floor coverings

▮ shower curtains

Write the name of the item that has the most plasticizer added to the PVC that it is made from.

_____

Give a reason for your choice.

_____   [2]

ii Rubber is used to make car tyres and elastic bands.

Which has more cross-linking: the rubber in car tyres or the rubber in elastic bands?

_____

Give a reason for your choice.

_____   [2]

Total [6]

1 Add the names of the missing electromagnetic radiations.

radio waves		infrared		ultraviolet		gamma rays

increasing energy ⟶

2 Write each of these electromagnetic radiations in the correct column of the table.

ultraviolet   **light**   **X-rays**   **infrared**
gamma rays   **microwaves**

Ionizing radiations	Radiations that cause a heating effect only

3 Use the words in the box to finish labelling the diagram.

> **detector**   **energy of one photon**   **intensity**
> **absorbs**   **number of photons**   **transmits**
> **reflects**   **source**

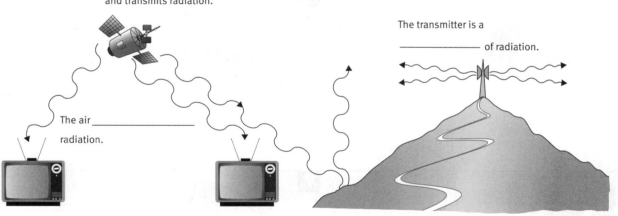

The satellite both _____ and transmits radiation.

The transmitter is a

_____ of radiation.

The air _____ radiation.

The TV is a _____ .

It _____ radiation.

The energy deposited here by a beam of radiation

depends on the _____

and the _____ .

The hill _____ radiation.

The energy that arrives at a surface each second

is the _____ of the radiation.

57

4 Add captions to each picture. Include

▶ the name of the type of electromagnetic radiation represented
  (radio waves, ultraviolet radiation, and so on)
▶ the damage (if any) this type of radiation can do to living cells
▶ what Alex can do to protect himself from this type of radiation
  (if he needs to do anything)

Alex's holiday: a day in the life

Picture 3 speech bubble: *Is that the dentist? My filling's just fallen out.*

5 Solve the anagrams in the box.

Then use the words to fill in the gaps.

camel chi	its ninety	i noizing	rat vibe	acne riots	emit

_____ radiation breaks particles into ions,

H which can then take part in other _____ _____.

Non-ionizing radiations can heat the materials they strike by

making their particles _____. The heating

effect depends on the radiation's _____

H and the length of _____ it strikes the material.

6 Write each letter in the correct section of the diagram.

A This gas is added to the atmosphere by respiration.

B This gas is made in the atmosphere when oxygen molecules absorb
ultraviolet radiation.

C This gas is removed from the atmosphere in a chemical process
that involves the absorption of ultraviolet radiation.

D This gas is present in the Earth's atmosphere.

E This gas helps to prevent humans getting skin cancer.

F Molecules of this gas contain at least two oxygen atoms.

G This gas is removed from the atmosphere by photosynthesis.

H This gas is added to the atmosphere by combustion.

I This gas absorbs radiation that the Earth emits, so keeping the Earth
warmer than it would otherwise be.

J This gas helps to prevent humans getting eye cataracts.

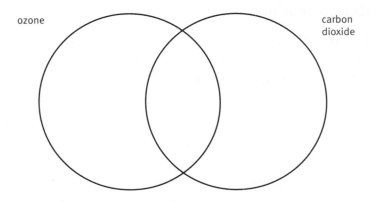
ozone   carbon dioxide

What is radiation?

Radiation

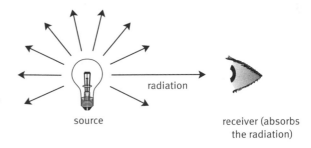

- carries energy
- spreads out (radiates) from its source
- may be detected by a receiver; receivers absorb the radiation

source

receiver (absorbs the radiation)

On its journey, radiation is reflected, transmitted, or absorbed by other materials.

How much energy does electromagnetic radiation deliver?

Electromagnetic radiation delivers energy in packets, or **photons**. The amount of energy delivered to a receiver depends on

- the number of photons that arrive
- how much energy each photon carries

The **intensity** of electromagnetic radiation is the energy that arrives at a surface in one second. The further the surface is from a source, the lower the intensity of radiation that hits it.

H This is because radiation spreads out as it gets further from the source.

What do different types of electromagnetic radiation do?

The electromagnetic spectrum includes these radiations:

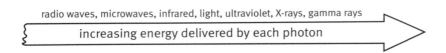

radio waves, microwaves, infrared, light, ultraviolet, X-rays, gamma rays

increasing energy delivered by each photon

Ultraviolet radiation, X-rays, and gamma rays are **ionizing radiations**. They break particles into ions that then can take part in other chemical reactions.

Radio waves, microwaves, light, and infrared radiation can **heat** materials by making their particles vibrate faster. The heating effect depends on the radiation's intensity
H and the length of time it strikes the material.

Some radiations are used to **transmit information**:

- Radio waves broadcast radio and TV programmes.
- Microwaves send messages between mobile phones and phone masts.
- Infrared radiation sends messages between remote controls and TVs.

What harm does electromagnetic radiation do?

▶ **Ionizing radiation:**
  – Large amounts kill cells.
  – Small amounts damage a cell's DNA. The cell may then grow uncontrollably to form a cancer tumour.

▶ The **heating effect** of absorbed radiation can damage cells.

▶ **Low-intensity microwave radiation** from mobile phone masts and handsets may be a health risk – scientists are not sure.

Barriers **protect** humans from radiation:

▶ Microwave ovens have metal cases that reflect microwaves to stop them leaving the oven.

▶ People wear clothes and sunscreen to absorb the Sun's ultraviolet radiation. These reduce the risk of getting sunburn or skin cancer.

How does electromagnetic radiation make life on Earth possible?

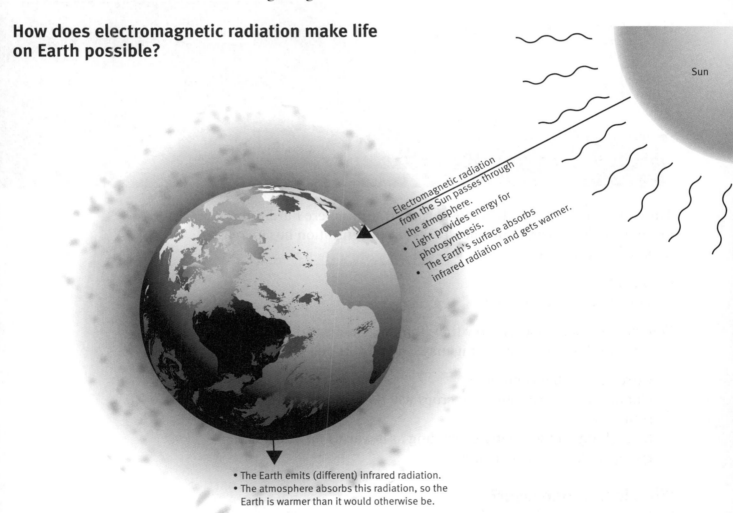

Sun

Electromagnetic radiation from the Sun passes through the atmosphere.
• Light provides energy for photosynthesis.
• The Earth's surface absorbs infrared radiation and gets warmer.

• The Earth emits (different) infrared radiation.
• The atmosphere absorbs this radiation, so the Earth is warmer than it would otherwise be.

This is the greenhouse effect.

What is global warming?

Greenhouse gases keep the Earth warmer than it would otherwise be.

There are three main greenhouse gases in the Earth's atmosphere:

- carbon dioxide (small amounts)
- methane (trace amounts)
- water vapour

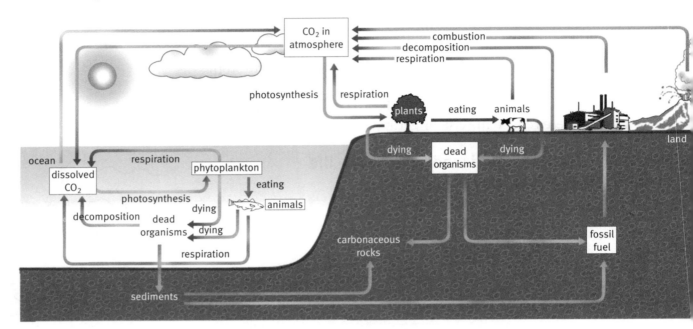

The concentration of carbon dioxide in the atmosphere hardly changed for thousands of years. But since 1800 its concentration has increased, mainly because humans

- burn fossil fuels for energy
- burn forests to clear land

Computer climate models show that human activities are causing global warming. Global warming may lead to

- extreme weather conditions
- climate change, so some food crops will no longer grow in some places
- ice melting and sea water expanding as it warms up, causing rising sea levels and flooding

What is the zone layer?

In the upper atmosphere, radiation acts on oxygen to make ozone. Ozone absorbs ultraviolet radiation and protects living organisms from its harmful effects.

1 a Draw straight lines to link each type of radiation with one
of its uses.

Type of radiation		Use
infrared		broadcasting television programmes
microwaves		transmitting messages between phone masts
radio waves		transmitting messages between remote controls and televisions

[3]

b Add to the diagram of a microwave oven by finishing the labels.

Choose words from the box.

emits	absorbs	transmits	reflects

source
_____
radiation

metal
_____
radiation

glass
_____
radiation

water in food
_____
radiation

[4]

c Name one feature of a microwave oven that protects its users
from the radiation it emits.

_____

_____

[1]

Total [8]

2 The Sun emits ultraviolet radiation.

   a Ultraviolet radiation can be harmful.

     Put ticks in the boxes next to the statements that are true.

       Ultraviolet radiation is a type of ionizing radiation. ☐

       Ultraviolet radiation makes molecules vibrate slower. ☐

       Ultraviolet radiation can make cells grow in an uncontrolled way. ☐

       Ultraviolet radiation makes molecules more likely to react chemically. ☐

       Ultraviolet radiation can damage the DNA of cells. ☐

       Ultraviolet radiation cools down molecules. ☐ **[2]**

   b Name one physical barrier that people use to protect themselves from the Sun's ultraviolet radiation.

     _____ **[1]**

   c **i** Put a tick in **one** box to show how the Earth's upper atmosphere protects living things from the Sun's ultraviolet radiation.

       Ozone molecules emit ultraviolet radiation. ☐

       Ozone molecules reflect ultraviolet radiation towards the Earth. ☐

       Ozone molecules transmit ultraviolet radiation in all directions. ☐

       Ozone molecules absorb ultraviolet radiation. ☐ **[1]**

     ii Some chemicals in aerosols destroy ozone molecules in the atmosphere.

      Explain why many governments have banned these chemicals.

      _____

      _____

      _____

      _____ **[2]**

Total [6]

3 Beams of electromagnetic radiation from the phone mast deliver photons ('packets') of energy to Mike's and Helen's mobile phones.

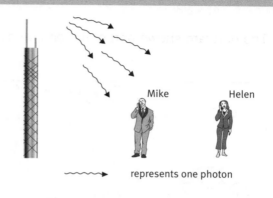

represents one photon

a The amount of energy delivered by each photon is the same. Explain why the amount of energy that arrives at Helen's mobile phone is less than the amount of energy that arrives at Mike's mobile phone.

_____ [1]

b The graph shows how the intensity of a beam of electromagnetic radiation changes as the distance from its source increases.

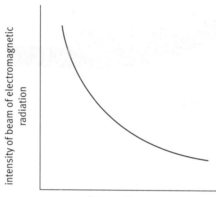

distance from source

i Complete this sentence:

As the distance from the source increases, the intensity

of the beam _____ .   [1]

ii Explain why the intensity of the beam changes in this way as the distance from the source increases.

_____

_____ [1]

c Complete these sentences. Choose words from the box.

intensity	vibrate	time	ionize	temperature

Electromagnetic radiation from the mobile phone makes Mike's

brain slightly warmer. This is because the radiation makes molecules

in the brain _____ faster. The size of the temperature

increase depends on the _____ of the radiation and

the length of _____ of his call.   [3]

Total [6]

4 The diagram shows part of the carbon cycle.

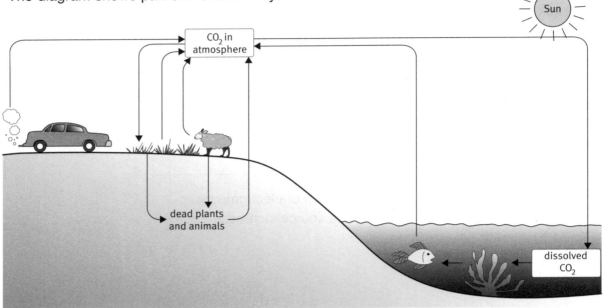

a i Name two processes that add carbon dioxide to the atmosphere.

_____ [2]

ii Name two processes that remove carbon dioxide from the atmosphere.

_____ [2]

b i Use the carbon cycle to explain why the amount of carbon dioxide in the Earth's atmosphere was approximately constant for thousands of years.

_____

_____

ii Give two reasons why the amount of carbon dioxide in the atmosphere has increased since 1800.

_____

_____ [3]

c Increasing amounts of carbon dioxide in the atmosphere cause global warming.

i Give two problems caused by global warming.

_____

_____ [2]

ii Name two greenhouse gases other than carbon dioxide.

_____ [2]

Total [11]

1 Decide which statements are about natural selection, which statements are about selective breeding, and which statements apply to both.

Write the letter of each statement in the correct part of the Venn diagram.

A Individuals within a species show variation.

B Humans choose individuals with the characteristics they want and breed from them.

C Some individuals have features that help them survive if the environment changes.

D The organisms that breed pass on their genes to their offspring.

E Individuals with features that help them to survive live longer and so are more likely to reproduce.

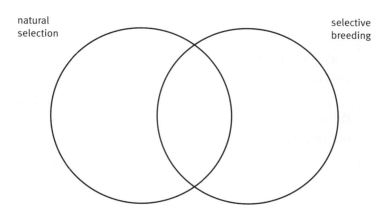

natural selection

selective breeding

2 Use the phrases in the box to label the diagram. You may use them more than once.

> **effector cells**   **receptor cells**   **neuron**   **central nervous system**

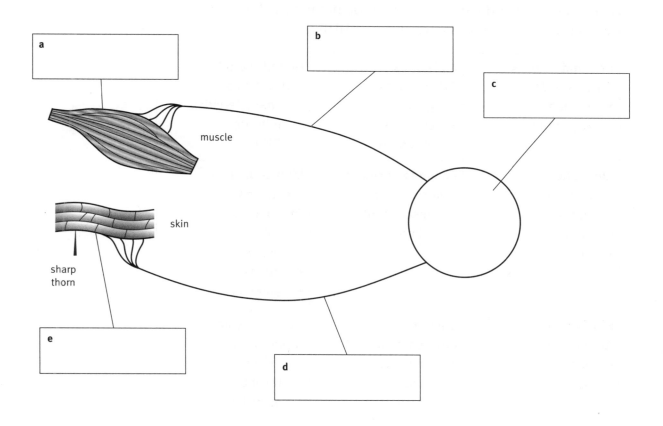

a

b

c

muscle

skin

sharp thorn

e

d

H **3** Make notes about homeostasis in the table.

> ▶ Write a title in the top row.
> ▶ Write the two or three most important points in the next row down.
> ▶ Write other, detailed, information in the lower rows.

Title:	
Most important points:	
Other information:	

4 Do this activity with a friend.
Define the word at the top of the card. Do not use the 'taboo' words.
Get your friend to guess the word you are defining.

Hominid species *Taboo words:* • human • big brains	**Mutations** *Taboo words:* • genes • changes • inherited	**Genetic variation** *Taboo words:* • characteristics • inherited • DNA
Competition *Taboo words:* • nutrients • water • survival	**Common ancestor** *Taboo words:* • descendants • related	**Multicellular organisms** *Taboo words:* • cells • many • plants • animals
Biodiversity *Taboo words:* • variety • species	**Sustainable development** *Taboo words:* • needs • future • environment	**Nervous system** *Taboo words:* • effector • receptor • electrical

5 Solve the clues to fill in the arrow-word.

Horizontal

3 The molecules that living things developed from were produced by environmental conditions on Earth at the time, or from elsewhere in the Solar . . .

7 All species of living things now on Earth . . . from simple living things.

10 Selective breeding . . . happens without the involvement of humans.

11 The molecules that living things developed from were produced by environmental conditions on . . . at the time, or from elsewhere in the Solar System.

13 The molecules that living things developed from were produced by environmental conditions on Earth at the time, or from elsewhere in the . . . System.

16 . . . is the symbol for the element oxygen.

Vertical

1 Nerve cells are also called . . .

2 Most scientists believe that life on Earth began . . . thousand five hundred million years ago.

3 Early humans with bigger brains had a better chance of . . . than those with smaller brains.

4 Mutated genes in . . . cells can be passed on to offspring.

5 Hormones . . . in the blood.

6 Living things depend on the environment and on other . . . of organisms for their survival.

8 . . . is the chemical that makes up chromosomes.

9 The first living things developed from molecules that could . . . themselves.

12 . . . is the symbol for the element hydrogen.

14 . . . is the symbol for the element carbon.

15 Changes that affect one species in a food web affect other species in the same . . . web.

How did life on Earth begin and evolve?

Life on Earth began about 3500 million years ago. Simple organisms developed from molecules that could copy themselves. Biologists disagree about the origin of these molecules – some believe that they were produced by environmental conditions on Earth at that time; others believe they came from elsewhere in the Solar System.

All species of living – and extinct – things on Earth evolved from the first simple organisms. Fossils and DNA analysis provide evidence for evolution.

What is evolution?

Evolution happens by **natural selection**. If conditions on Earth had been different, evolution would have happened differently. The number and variety of species that now exist may have been different.

The diagram shows how natural selection works. It is different from selective breeding, in which humans choose characteristics they want individuals to have.

Variation between individuals of a species is caused by both the **environment** and **genes**. Only genetic variation can be passed from one generation to the next, so without genetic variation there would be no natural selection.

H Three things make genes change, or **mutate**:

- mistakes when copying chromosomes
- ionizing radiation
- some chemicals

If sex cell genes mutate, three things may happen:

- The mutation may have no effect.
- The fertilized egg may not develop.
- The offspring may have a better chance of surviving and reproducing. Then the mutated gene passes on to the next generation and becomes more common.

Over many years and generations, **new species** have evolved.

H This is the result of the effects of mutations, environmental change, and natural selection.

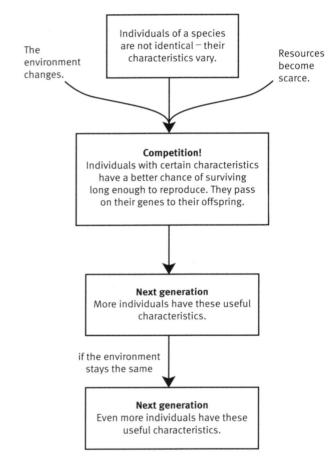

The environment changes.

Individuals of a species are not identical – their characteristics vary.

Resources become scarce.

Competition!
Individuals with certain characteristics have a better chance of surviving long enough to reproduce. They pass on their genes to their offspring.

Next generation
More individuals have these useful characteristics.

if the environment stays the same

Next generation
Even more individuals have these useful characteristics.

How did humans evolve?

Hominids are animals that are more like humans than apes. Many different hominid species evolved from a common ancestor. Those with **bigger brains** and who **walked upright** had a better chance of surviving. Gradually all hominid species, except for modern humans (*Homo sapiens*), became extinct.

How are human communication systems organised?

In living things with many cells – **multicellular organisms** – cells are specialized for different jobs. Multicellular organisms evolved two communication systems.

Nervous system

The nervous system uses electrical impulses to transmit messages and respond to them **quickly**. For example, you cough if you breathe in smoke and you blink if a fly gets in your eye.

In vertebrates, the nervous system is made of nerve cells (neurons). It is controlled by the central nervous system (the brain and spinal cord).

Hormones

Hormones are **chemicals** that travel in the blood. They carry information all over the body and bring about long-lasting responses. Hormones transmit messages more **slowly** than nerves.

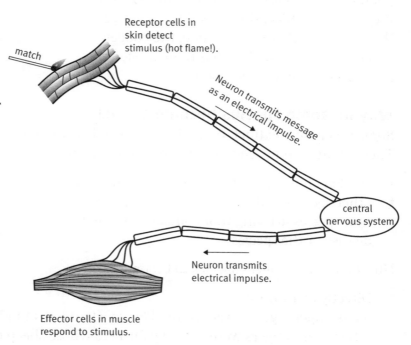

Receptor cells in skin detect stimulus (hot flame!).

match

Neuron transmits message as an electrical impulse.

central nervous system

Neuron transmits electrical impulse.

Effector cells in muscle respond to stimulus.

Testosterone is a hormone. At puberty, it causes many changes in boys: the testes start to make sperm, the voice deepens, and pubic hair grows.

Insulin is another hormone. After a high-carbohydrate meal, the pancreas detects a high concentration of glucose in the blood. The pancreas makes insulin. Insulin causes the liver to remove glucose from the blood.

Homeostasis

Both the nervous and hormonal communication systems help to keep a constant internal environment in organisms. This is **homeostasis**.

For example, if you are too hot, the brain detects that the temperature of the blood flowing through it is too high. It sends an electrical impulse through the nervous system to your sweat glands. You start to sweat more.

Interdependence of organisms

Living organisms depend on **other species** and the **environment** for their needs, including nutrition and shelter. Within one habitat, species compete for resources. For example, in gardens, weeds compete with vegetables for light, water, nutrients, and space.

A food web shows how animals that live in a particular habitat meet their nutritional needs. It shows what eats what.

Changes to one part of the food web affect other species in the same food web. For example, many foxes may get ill and die. The fox population decreases. The populations of foxes' prey species – mice, slugs, beetles, and frogs – then increase. At the same time, the badger population increases because there are fewer foxes to compete with for food.

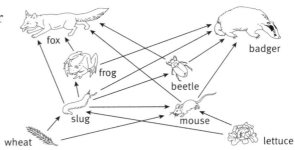

Why do some species become extinct?

Some species are extinct, or are in danger of becoming extinct. These changes may lead to a species becoming extinct:

▶ sudden changes in environmental conditions
▶ other species in the food web becoming extinct
▶ new species arriving that compete with, eat, or cause disease of the species

Human activity causes species to become extinct:

▶ Directly by hunting:
  – Passenger pigeons were hunted for sport and meat in North America.
  – Tasmanian tigers were killed by farmers whose sheep they ate.
▶ Indirectly by pollution or destroying habitats:
  – A cow medicine has poisoned and killed most vultures in India.
  – Pandas are endangered because their habitat is scarce.

Why is biodiversity important?

There is enormous variety in living organisms. This is called **biodiversity**. Biodiversity is important for many reasons:

▶ We may develop new medicines from species that we do not yet use.
▶ For new food sources: – breeding cultivated rice with wild species may produce rice that is resistant to disease.

Biodiversity is an important part of **sustainable development**: using the environment to meet the needs of people today without damaging Earth for the people of the future.

1 This question is about communication systems in humans.

a Complete the following sentences.

Choose words from this list:

electrical   long   short   slowly   quickly   chemicals   brain

The nervous system uses _____ impulses to transmit

messages. It is controlled by the _____ and spinal

cord. Hormones are _____ that travel in the blood.

They transmit messages more _____ than nerves.

Hormones bring about more _____-term responses

than the nervous system.   [5]

b Look at these examples of human responses.
Some are brought about by the nervous system, others
by hormones.

Write the letters A, B, C, D, E, and F in the correct
columns of the table.

A coughing when you breathe in smoke

B blinking when a fly gets in your eye

C controlling blood sugar levels

D developing breasts at puberty

E controlling how quickly you grow taller

F moving away quickly if you touch something hot

Changes brought about by the nervous system	Changes brought about by hormones

[3]

Total [8]

2 Scientists have studied how 40 species of the cat family evolved. They discovered that lions and domestic cats shared a common ancestor 10.8 million years ago.

   a How might the scientists have obtained evidence to support their explanation?

      Tick the two best answers.

        studying fossils ☐

        analysing the fur of cat ancestors ☐

        analysing the DNA of modern cats, lions, and other species of the cat family ☐

        analysing the blood of cat ancestors ☐   **[2]**

   b **i** Complete this sentence.

        Domestic cats and lions evolved as a result of a process

        called natural _____.   **[1]**

     ii The stages below explain how evolution made changes to one species: lions.

        The stages are in the wrong order.

        A More individuals in this generation had features that helped them survive in their new environment.

        B Early lions migrated from Asia to Africa. Some individuals had features that helped them survive in the new environment.

        C Individual lions are not identical; the species shows variation.

        D These lions bred. They passed on their genes to their cubs.

        Fill in the boxes to show the correct order. The first one has been done for you.

C			

    [3]

  H **c** The diagram shows scientists' ideas about when some species of the cat family began evolving from their common ancestor. For example, the domestic cat and the ocelot last shared a common ancestor 8.0 million years ago.

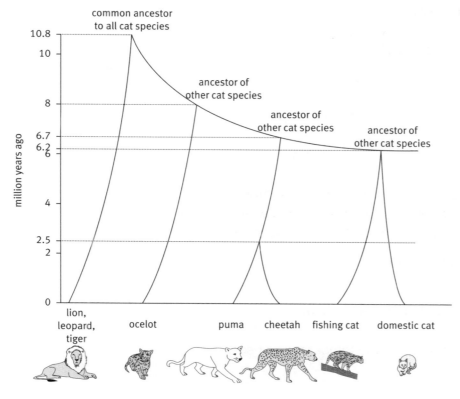

i   When did the cheetah and the domestic cat last share a common ancestor?

_____   [1]

ii Which species on the chart probably has DNA that is most similar to that of the domestic cat?

_____   [1]

iii Read the information in the box.

> The common ancestor of the puma and cheetah lived in North America. Individuals of the ancestor species migrated. Some went to South America and evolved into a new species: the puma. Others went to Africa and evolved to become a different species: the cheetah.

Complete the following sentences. Choose words from this list.

| **mutations** | **survival** | **selection** | **environmental** |

This shows how _____ changes help to produce

new species. Changes to genes (_____) and natural

_____ also help to produce new species.   [3]

Total [11]

3 Read the information in the box.

Here is part of an Antarctic food web:

Scientists have discovered that the sea temperature around Antarctica has risen by 1°C since 1960. Warmer sea water creates problems for animals that live on the seabed. For example,

▶ scallops become unable to swim
▶ limpets cannot turn over

This makes it easier for predators to catch them.

a What is most likely to have made the temperature of sea water increase?

Draw a ring round the best answer.

acid rain   global warming   the thinning ozone layer   [1]

b i Use the food web to name one predator of the Antarctic scallop.

_____

ii If the Antarctic scallop population decreases, what is likely to happen to the populations of its predators?

_____   [2]

c Scientists fear that if Antarctic sea temperatures continue to rise, some species may become extinct.

Use the information in the box above to tick the **two most likely reasons** for the possible future extinction of the brittlestar.

Environmental conditions change. ☐

A new species that is a competitor of the brittlestar is introduced to Antarctica. ☐

Another living thing in the brittlestar's food chain becomes extinct. ☐

A new species that is a predator of the brittlestar is introduced to Antarctica. ☐   [2]

Total [5]

1 Make up seven sentences using the phrases in the table.
Each sentence must include a phrase from each column.
Write your answers in the grid at the bottom.

For example, the sentence
Carbohydrates are natural polymers
becomes X1A.

X Carbohydrates	**1** are	**A** natural polymers.
	2 are made from	**B** muscle, tendons, skin, hair, and haemoglobin.
	3 include	**C** the elements carbon, hydrogen, oxygen, and nitrogen.
Y Proteins		**D** the elements carbon, hydrogen, and oxygen.
	4 are broken down by digestive enzymes to make small soluble molecules of	**E** amino acids.
		F cellulose, starch, and sugars.
	5 are the main constituents of	**G** glucose sugar.

X							
1							
A							

2 Match each **type of additive** to a **purpose** for adding it to food.

Type of additive
colouring
artificial sweeteners, for example aspartame
flavour enhancers, for example monosodium glutamate

Purpose for adding this additive to food
to make a food taste better
to make a food look attractive
to make a food taste sweeter without adding sugar

3 The information in the table is about harmful chemicals in food.

Fill in the empty boxes.

Type of chemical	Health concern	Examples of foods affected
natural chemicals in plants		some types of mushroom
		cassava
	some people are allergic to them	
aflatoxins produced by mould during crop storage	may increase risk of getting cancer and damage the liver	
chemicals that form during food processing and cooking	some, e.g. acrylamide, increase the risk of getting cancer	

4 Cross out the words that are wrong.

There are two types of diabetes. Type 1 diabetes begins suddenly, when a child's **pancreas/liver** stops making enough **glucose/insulin**. People with type 1 diabetes need regular insulin injections.

Type 2 diabetes usually starts in **children/teenagers/adults**, when body cells stop responding to **growth hormone/glucose/insulin**. It is linked to a diet **low/high** in **fat/protein** and to obesity. People with type 2 diabetes must control their diet carefully and exercise sensibly. They sometimes need insulin injections.

5 Here's what happens to protein in your body.
Write a number in each box to show a sensible order
for the cartoon strip.

The cartoon is not drawn to
scale – all molecules are
shown greatly magnified.

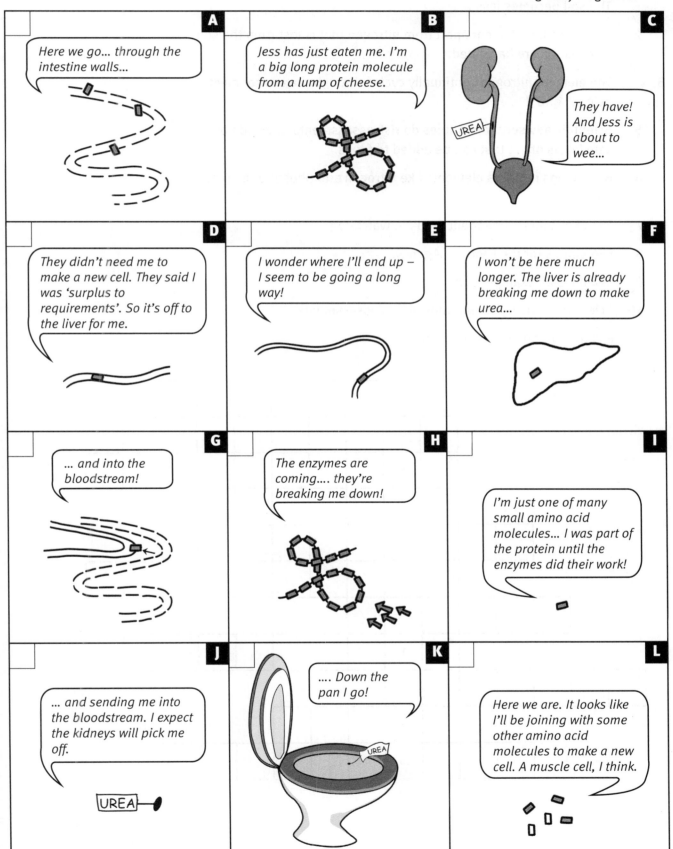

A Here we go... through the intestine walls...

B Jess has just eaten me. I'm a big long protein molecule from a lump of cheese.

C They have! And Jess is about to wee...
UREA

D They didn't need me to make a new cell. They said I was 'surplus to requirements'. So it's off to the liver for me.

E I wonder where I'll end up – I seem to be going a long way!

F I won't be here much longer. The liver is already breaking me down to make urea...

G ... and into the bloodstream!

H The enzymes are coming.... they're breaking me down!

I I'm just one of many small amino acid molecules... I was part of the protein until the enzymes did their work!

J ... and sending me into the bloodstream. I expect the kidneys will pick me off.
UREA

K .... Down the pan I go!
UREA

L Here we are. It looks like I'll be joining with some other amino acid molecules to make a new cell. A muscle cell, I think.

6 Solve the clues to fill in the grid.

1 When crops are harvested, the soil loses elements like nitrogen. The soil becomes less . . .

2 . . . is another element apart from nitrogen that is lost from the soil when crops are harvested.

3, 4 The element nitrogen continually cycles through the environment by c . . . and d . . .

5 Scientific advisory committees do risk assessments to decide on safe levels of . . . that can be added to foods.

6 H . . . crops removes elements like nitrogen and phosphorus from the soil.

7 The FSA is an independent food . . . watchdog.

8 FSA stands for the Food . . . Agency.

9 Insulin is an example of a . . .

10 Scientific advisory committees do . . . assessments.

11 P is the symbol for the element . . .

What chemicals are living things made from?

Many chemicals in living things are **natural polymers**. Two important natural polymers are

- **Carbohydrates**, for example cellulose, sugars, and starch. These are made from the elements carbon, hydrogen, and oxygen.
- **Proteins**, which are amino acids joined together in long chains. They are made mainly from carbon, hydrogen, oxygen, and nitrogen.

How do elements cycle through living organisms?

Living things take in chemicals by photosynthesis, breathing, absorption, and eating. Decomposers, like microbes and fungi, break down dead plants and animals. These processes recycle the materials in living things.

There are nitrogen atoms in the protein molecules of all cells. These atoms are used over and over again by different organisms. The continual cycling of nitrogen compounds is the **nitrogen cycle**.

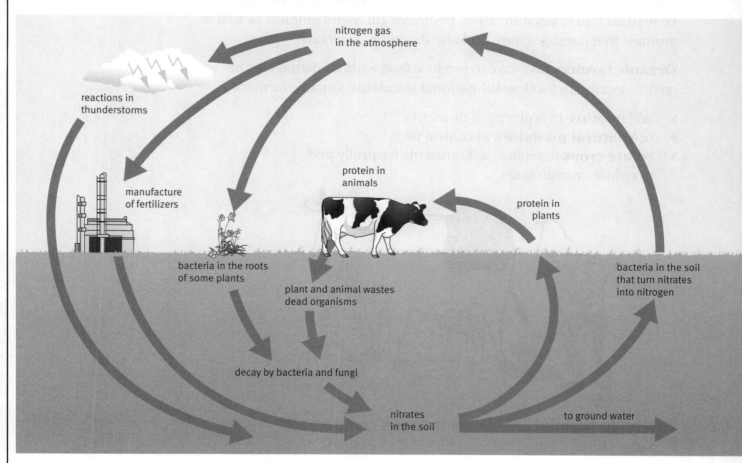

Other elements, like phosphorus and potassium, are also continually recycled.

How are intensive and organic farming different?

Farmers use the same land every year to grow food crops. When farmers harvest crops, the soil loses nitrogen, phosphorus, and potassium compounds. Farmers must replace these elements; if they do not, soil fertility decreases and the land produces less food. Pests and diseases also reduce crop yields.

Intensive farmers produce as much food from their land as possible. They aim to maximize yields by

▶ adding **synthetic fertilizers** to replace soil nutrients
▶ adding **pesticides** and **fungicides** to kill pests and disease-causing fungi
▶ adding **herbicides** to kill weeds that compete with crops for nutrients, light, and water

Organic farmers believe that intensive farming is unsustainable. Synthetic fertilizers do not recycle nutrients. Nitrates from fertilizers may be washed into rivers if misused. Pesticides kill useful animals as well as animals that damage crops, and also disrupt food chains.

Organic farmers take care to produce food without damaging the environment. They follow UK national standards. Organic farmers

▶ add **manure** to replace soil nutrients
▶ use **natural predators** to control pests
▶ **rotate crops** to replace soil nutrients naturally and to reduce crop diseases

Where do harmful chemicals in foods come from?

▶ Some plants contain harmful **natural chemicals**, for example
  – poisonous mushrooms
  – under-processed cassava
  – gluten in flour (some people are intolerant to this)
▶ **Moulds**, like aflatoxin, occasionally contaminate stored cereals.
▶ Traces of **pesticides** and **herbicides** may remain in foods.
▶ Harmful chemicals form during food **processing** and **cooking**. For example acrylamide, which may increase the risk of cancer, is made in crisp manufacture.

Why do manufacturers add chemicals to food?

Additive type	Purpose of additive
colourings	to make food look attractive
flavourings	to improve taste
sweeteners	to make food taste sweeter without adding sugar. Sweeteners such as saccharin are many times sweeter than sugar so only very small amonnts are needed.
emulsifiers	to mix together ingredients that normally separate, like oil and water
preservatives	to prevent harmful microbes growing and to keep food safe for longer
antioxidants	to prevent fats and oils deteriorating (going rancid) by reacting with oxygen in the air

Food additives with E numbers have passed safety tests. They have been approved for use in the European Union. Scientific advisory committees do risk assessments to decide on safe levels of additives.

How can we avoid foods that contain harmful chemicals?

Food labels help consumers decide what to buy and to avoid eating harmful chemicals. They give information about ingredients, including additives and nutrients. But labels do not always mention health risks, and may be misleading. For example a food that 'contains 50% less fat' may still have a high fat content.

The UK's independent food safety watchdog is the **Food Standards Agency** (FSA). It aims to make sure that food is safe, healthy, and fairly marketed.

How do we digest food?

Enzymes in the digestive system help break down natural polymers into small, soluble compounds. These pass through the gut wall and dissolve in the blood. The blood then transports them around the body.

Big **protein** molecules break down into smaller **amino acid** molecules.

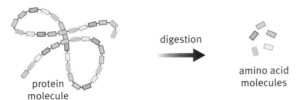

Big **starch** molecules break down into small **glucose** molecules.

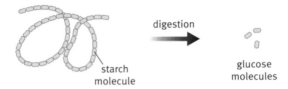

The blood transports amino acids to cells. Cells grow when amino acids join together to make new proteins. Muscles, tendons, skin, hair, and haemoglobin (in the blood) are all made mainly from protein.

If the body has more amino acid molecules that it needs, it sends them to the liver. The liver breaks them down into poisonous urea. The bloodstream transports urea to the kidneys. The kidneys remove urea from the blood and excrete it in urine.

What is diabetes?

Glucose is quickly absorbed into the bloodstream, so eating foods high in glucose causes a rapid rise in blood sugar level. In most people, the hormone **insulin** controls blood sugar level. The bodies of people with diabetes do not control blood sugar levels properly.

There are two types of diabetes: type 1 and type 2.

Type	1	2
How it starts	Starts suddenly in childhood when immune system cells attack cells in the pancreas.	Usually starts in adulthood. Linked to ▶ poor diet ▶ obesity ▶ genetics ▶ age
What happens	The pancreas stops producing enough insulin.	The body stops responding to its own insulin or does not make enough insulin.
How to control it	insulin injections	diet and exercise

1 **a** Draw straight lines to link each type of **food additive** to
a **reason** why it is added to food.

Type of additive		Reason
flavourings		to stop harmful microbes growing
emulsifiers		to make food taste better
preservatives		to prevent fats and oils reacting with oxygen from the air
antioxidants		to mix ingredients together that would normally separate

[3]

b Many food additives have E numbers.

Which statements about E numbers are true?

Put a tick in each of the **two** correct boxes.

There are definitely no health risks associated with
any additive that has an E number.   ☐

Food additives with E numbers have passed safety
tests.   ☐

Emulsifiers are the only food additives that have
E numbers.   ☐

Food additives with E numbers have been approved
for use in the European Union.   ☐

Food additives with E numbers have been approved
for use in England only.   ☐   [2]

Total [5]

2 The table shows the amount of fat in three products from a burger restaurant.

Food	Amount of fat per serving (grams)
large cheeseburger	27
regular chips	16
large milkshake	13

The bar chart shows the recommended maximum daily fat intake for different groups of people.

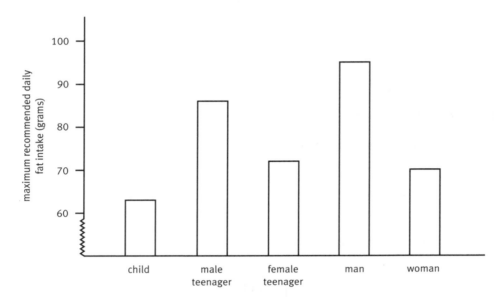

a Finish the sentence.

Choose a phrase from the list.

more than   **less than**   **the same as**

The recommended daily fat intake for adults is _____

the recommended daily intake for a child.

[1]

b Kezi is a 15-year-old girl. For lunch she has

▶ a large cheeseburger
▶ regular chips
▶ a large milkshake

i Calculate the total amount of fat in Kezi's lunch.

Answer = _____ g

[1]

ii How much more fat could Kezi eat before she went over the recommended daily fat intake?

_____

[1]

c During digestion, enzymes help to break down big molecules into smaller molecules.

Why must big molecules be broken down into smaller molecules?

Put a tick in each of the two correct boxes.

Big molecules dissolve in the blood more easily than small molecules. ☐

Small molecules dissolve in the blood more easily than big molecules. ☐

Only small molecules can pass through the intestine walls. ☐

Only big molecules are absorbed through the intestine walls. ☐ [2]

d Milkshake is high in protein.

Name four elements present in proteins.

_____ [2]

Total [7]

3 a i Give one reason why farmers protect their crops against pests and diseases.

_____ [1]

ii Organic and intensive farmers protect their crops against pests and diseases in different ways.

Put ticks in the boxes next to each method that **organic** farmers might use to protect their crops.

Use natural predators to eat pests. ☐

Use synthetic pesticides. ☐

Rotate their crops regularly (grow a different crop every year in a certain field). ☐

Spray their crops with synthetic fungicides. ☐ [2]

b Look at the nitrogen cycle.

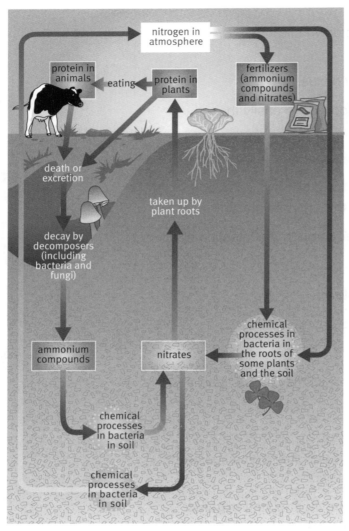

i Give two ways in which nitrates are removed from the soil.

_____

_____ [2]

ii How do nitrates spread through the soil?

_____ [1]

c Pat and Tony are organic farmers.

Give two things they can do to make sure there are enough nitrogen compounds in the soil for their crops to grow well.

_____

_____ [2]

Total [8]

1 Draw arrows to link each label to one or both people.

A This person has breathed in a radioactive chemical. He is contaminated.

D Big doses of ionizing radiation kill cells.

B A radioactive source is irradiating this person.

E Smaller doses of ionizing radiation can damage the DNA in cells.

C Ionizing radiation will stop hitting body cells
  ▶ either when the radioactivity of the source decreases to zero
  ▶ or when the source is removed from body

F Ionizing radiation stops hitting body cells when he moves away from the source.

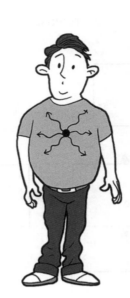

2 Solve the anagrams.

Use your answers to fill in the gaps. Use one word twice.

iznogini	mad age	illk
run dog	paces	chill my ace
stoma	men tatter	mad lice

We are exposed to _____ radiation all the time. Natural

sources of this radiation include radon gas from the _____

and cosmic rays from _____. Food and drink also

contain radioactive _____. Some people are exposed to

this type of radiation during _____ _____

or at work.

When _____ radiation hits cells it changes them and

makes them more likely to react _____.

Big doses _____ cells. Smaller doses _____

the DNA in cells.

3   Write each letter in an appropriate box to show how to generate electricity.

A   heat up water to make steam

B   wave movement

C   tidal movement

D   generator – a big coil of wire turns in a magnetic field

E   solar voltaic cells

F   releases carbon dioxide gas

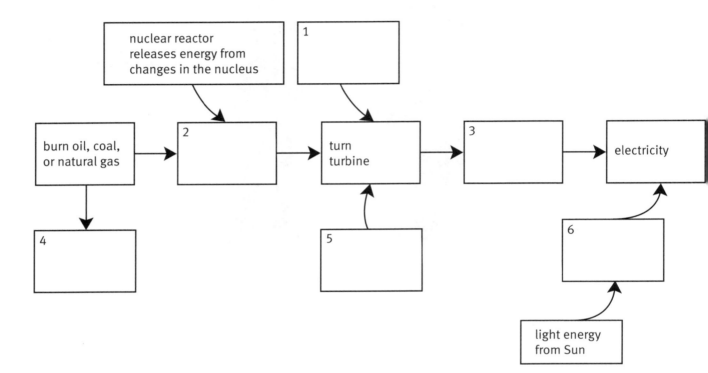

4   The data in the table is about electricity generated from different energy sources. Use the data to write one argument that supports each person's opinion.

	Nuclear	Wind	Coal
Approximate efficiency	35%	59% of wind's energy can be extracted by blades	35%
Environmental impact	produce radioactive waste	some people think they look unattractive	contribute to acid rain
Cost per unit of electricity (pence)	3.0–4.0	1.5–2.5	3.0–3.5
Tonnes of carbon dioxide made for one terajoule of electricity	30	10	260

Opinions:

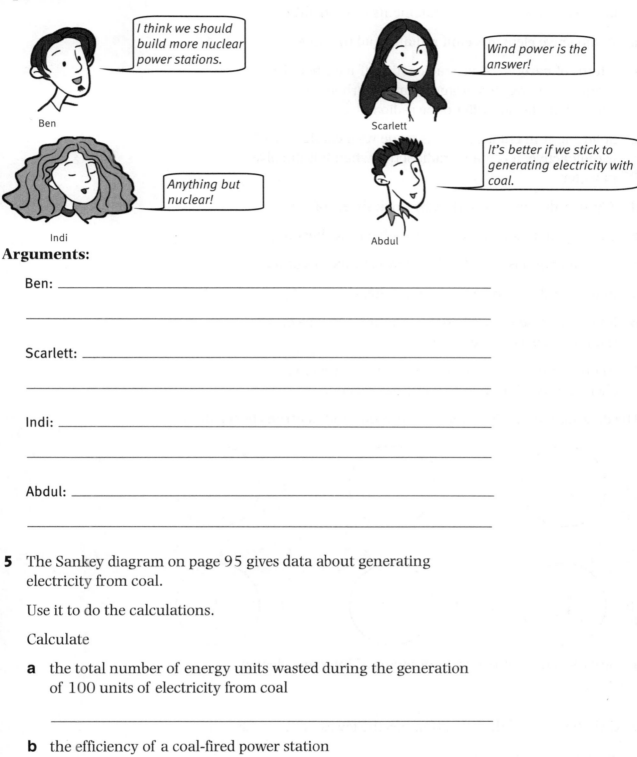

Ben

I think we should build more nuclear power stations.

Scarlett

Wind power is the answer!

Indi

Anything but nuclear!

Abdul

It's better if we stick to generating electricity with coal.

Arguments:

Ben: _____

_____

Scarlett: _____

_____

Indi: _____

_____

Abdul: _____

_____

5 The Sankey diagram on page 95 gives data about generating electricity from coal.

Use it to do the calculations.

Calculate

a the total number of energy units wasted during the generation of 100 units of electricity from coal

_____

b the efficiency of a coal-fired power station

_____

6 Write the letter **T** next to the statements that are true.
Write the letter **F** next to the statements that are false.

a Radioactive elements emit radiation all the time. _____

b Atoms of carbon-14 are radioactive. If a carbon-14
atom joins to oxygen atoms to make carbon dioxide,
the carbon dioxide will not be radioactive. _____

c Solid caesium chloride that is made with caesium-137
is radioactive. It remains radioactive when it is dissolved
in water. _____

d Alpha radiation is absorbed by thick sheets of lead. _____

e Gamma radiation can pass through a thin sheet of paper. _____

f Beta radiation is absorbed by a few centimetres of air. _____

g Radiation dose is measured in half-lives. _____

h Radiation dose is based on the amount and type of
radiation a person is exposed to. _____

i About a million times more energy is released in a
chemical reaction than in a nuclear reaction. _____

H **7** The diagrams show the number of protons and neutrons in six atoms.

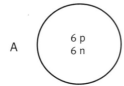

A 6 p 6 n

B 6 p 8 n

C 7 p 7 n

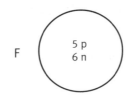

D 8 p 8 n

E 8 p 10 n

F 5 p 6 n

a Give the letters of two pairs of atoms of the same element.

_____

b Give the letter of the atom that has the fewest total number of

particles in its nucleus. _____

c Give the letter of the atom that has the greatest total number of

particles in its nucleus. _____

d Give the letters of two atoms that have the same total number of

particles in their nuclei. _____

What are radioactive materials?

Some materials give out (emit) ionizing radiation all the time. You cannot change the behaviour of a radioactive material – it emits radiation whatever its state (solid, liquid, gas) and whether or not it takes part in a chemical reaction. Most radioactive materials are found naturally in the environment.

What types of ionizing radiation are there?

Type of radiation	What it is	Penetration properties
alpha	a particle	absorbed by paper or a few cm of air
beta	a particle	penetrates paper; absorbed by a thin sheet of metal
gamma	a high-energy electromagnetic wave	absorbed only by ▶ thick sheets of dense metals, e.g. lead ▶ several metres of concrete

Why are some materials radioactive?

Every atom has a tiny core, or **nucleus**. The nucleus is surrounded by **electrons**.

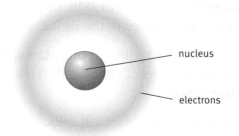

The nucleus is made of **protons** and **neutrons**. Every atom of a certain element has the same number of protons. Different atoms of this element may have different numbers of neutrons. For example:

Name	Number of protons	Number of neutrons	Is this type of carbon radioactive?
carbon-12	6	6	no
carbon-14	6	8	yes

The nucleus of a carbon-12 atom is stable. It is not radioactive.

The nucleus of a carbon-14 atom is unstable. It decays to make a stable nucleus of another element, nitrogen-14. As it decays, it emits beta and gamma radiation.

Other elements have atoms with unstable nuclei, for example radium-226. When radium-226 decays, it emits alpha and gamma radiation to make an atom of radon-222.

How does a material's radioactivity change with time?

As a radioactive material decays, it contains fewer and fewer atoms with unstable nuclei. It becomes less radioactive and emits less radiation. The time taken for the radioactivity to fall to half its original value is the material's **half-life**. Different radioactive elements have different half-lives. For example:

Radioactive element	Half-life
plutonium-242	380 thousand years
carbon-14	5.6 thousand years
strontium-90	28 years
iodine-131	8 days
lawrencium-257	8 seconds

What are the risks from radioactive sources?

We are exposed to **background radiation** all the time. Background radiation sources include:

▶ radon gas from the ground
▶ cosmic rays
▶ food and drink
▶ gamma rays from the ground and buildings

If ionizing radiation reaches you, you are **irradiated**. If a radioactive material gets onto your skin, or inside your body, you are **contaminated**. You will be exposed to the radiation as long as the material stays there.

When ionizing radiation hits atoms, it changes them. The atoms are more likely to react chemically. This is why ionizing radiation damages **living cells**:

▶ Big radiation doses kill cells.
▶ Smaller radiation doses can damage a cell's DNA. The cell may grow uncontrollably and cancer develops.

How is ionizing radiation useful?

▶ To **treat cancer**. In one type of **radiotherapy**, doctors put metal wires containing radioactive materials into the patient, near the tumour. The ionizing radiation damages cancer cells and they stop growing. However, the radiation also damages healthy cells.
▶ Gamma radiation kills bacteria on food and medical equipment, so **sterilizing** them.

How are radioactive materials handled safely?

The more ionizing radiation that a person is exposed to, the greater the risk of cancer. Some people are exposed to radioactive sources at work, for example staff in nuclear power stations and in hospital radiotherapy departments. Their exposure is monitored carefully.

Radiation dose measures the possible harm to your body. It takes account of the amount and type of radiation. Its units are sieverts (Sv).

How is electricity generated?

Electricity is easy to use and to transmit over long distances. It is a **secondary energy source** – it must be generated from another energy source, like gas.

Electricity is generated by these steps:

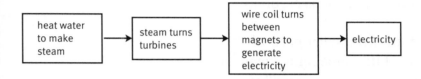

Generating and distributing electricity is not 100% efficient. At each step, energy is dissipated by heating the surroundings.

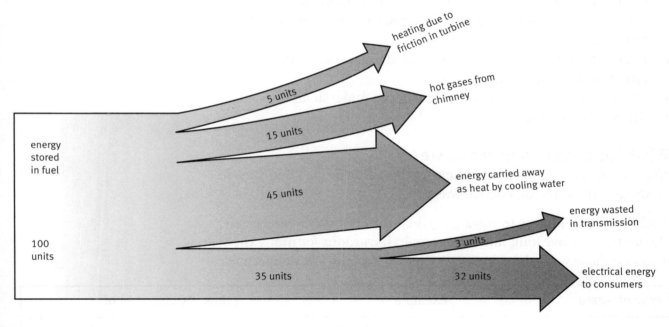

What are the energy sources for electricity generation?

▶ **Coal or gas**: burning coal or gas heats water to make steam. Steam turns the turbine. Coal- and gas-fired power stations make waste carbon dioxide.

▶ Generating electricity from **renewable energy sources** does not make carbon dioxide. These energy resources will never run out. Three examples are
  – Moving air turns **wind** turbines.
  – **Wave** movements turn turbines.
  – **Solar power** – photovoltaic cells generate electricity from the Sun's radiation.

Ⓗ How do nuclear power stations generate electricity?

Nuclear fuels release energy from changes in the nucleus.

They are used to generate electricity like this:

▶ **Fuel rods** in the nuclear reactor contain uranium-235.
▶ Neutrons are fired at the U-235.
▶ When a neutron hits a U-235 nucleus, the nucleus becomes unstable.
▶ The unstable U-235 nucleus splits into two smaller parts of about the same size – this is **fission**. At the same time, the nucleus releases more neutrons. Energy is released – about a million times more than in a chemical reaction.
▶ These neutrons hit more U-235 nuclei. Fission happens again. A **chain reaction** has started.
▶ **Control rods** absorb neutrons. They control the rate of fission reactions when they are lowered into or raised out of the reactor.
▶ Energy from the fuel rods is transferred as heat to a **coolant** (water or carbon dioxide).
▶ The hot coolant heats water in a boiler to make steam.
▶ The steam turns a turbine.

What happens to nuclear waste?

Nuclear power stations do not make carbon dioxide gas. But they do produce dangerous radioactive waste.

Scientists use half-lives to work out when nuclear waste will be safe. Elements that have long half-lives remain hazardous for many thousands of years; those with short half-lives quickly become less dangerous.

Type of waste	Example	How it's disposed of
low level	used protective clothing	packed in drums and dumped in a lined landfill site
intermediate level	materials that have been inside reactors – may remain dangerously radioactive for many years	mixed with concrete and stored in stainless steel containers
high level	concentrated radioactive material – decays fast and releases energy rapidly so needs cooling	difficult to store safely as radiation damages container; chemically corrosive

1   Read the article about treating cancer with radioactive materials.

> Arthur has a cancer tumour deep inside his body. His doctors will use radiotherapy to treat it. Arthur's doctors and radiotherapists plan the treatment carefully. They tattoo his skin to show exactly where to direct the radiation, and calculate the dose of radiation Arthur must receive.
>
> Arthur gets his treatment in a lead-lined room. When everything is ready, the radiotherapist leaves the room. Once outside, she switches on the treatment machine. Gamma rays enter Arthur's body for a few minutes. During the treatment, the radiotherapist watches Arthur on closed-circuit television. They can talk to each other over an intercom.
>
> Arthur goes to hospital for treatment every weekday for five weeks. On each visit, the gamma radiation enters his body at a different angle.

a   i   What type of material emits the radiation that enters Arthur's body?

_____

ii   Why does the radiotherapist use gamma radiation, and not alpha or beta radiation?

_____

_____   [2]

b   i   What does gamma radiation do to cancer cells?

_____

ii   Why is it important to direct the radiation exactly at the cancer tumour?

_____

_____   [2]

c   i   Why are the walls of the treatment room lined with lead?

_____

_____

ii   Why does the radiotherapist leave the room while Arthur is receiving his treatment?

_____

_____   [2]

Total [6]

2 Caesium-137 (Cs-137) emits beta particles. It is used to treat some cancers.
The graph shows how the activity of this radioactive source changes over time.

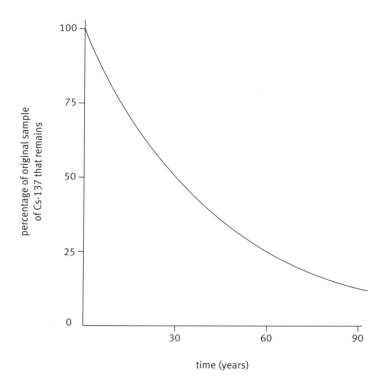

time (years)

a Read the statements below.
Put ticks in the boxes next to each true statement.

The activity of the Cs-137 source decreases over time. ☐

Most radioactive elements have a half-life of between 10 and 50 years. ☐

The half-life of Cs-137 is 30 years. ☐

The longer the half-life of a radioactive source, the more quickly it becomes safe. ☐

Beta radiation is absorbed only by thick sheets of lead or concrete. ☐

[2]

b A sample of caesium chloride contains 10 g of caesium-137.

Calculate the mass of caesium-137 that will remain after 60 years.

_____

_____

[2]

c Caesium-137 decays to barium-137. Barium-137 is not radioactive.

Complete the following sentences.

Use the words in the box.

| negative | unstable | stable | neutral |

The nucleus of a caesium-137 atom is _____. It decays

and emits beta radiation. This makes barium-137, which has a nucleus

that is _____.   [2]

Total [6]

3 a Nuclear power stations generate electricity.
The stages in this process are shown below.

A These neutrons hit more uranium-235 nuclei. Fission happens
again. A chain reaction has started.

B The steam turns a turbine.

C Energy from the fission reaction is transferred as heat to
a coolant, such as water or carbon dioxide.

D The unstable nucleus splits into two smaller parts of about
the same size. This is fission. At the same time, the nucleus
releases more neutrons.

E Neutrons are fired at fuel rods.

F When a neutron hits the nucleus of a uranium-235 atom,
the nucleus becomes unstable.

G The hot coolant heats up water in a boiler to make steam.

The stages are in the wrong order.

Write a letter in each box to show the correct order.

| E | F | | | | | |

[4]

H

b Complete the following sentences.
Choose from the words in the box.

barium	protons	electrons	boron
bismuth	neutrons	rate	

Control rods control the _____ of fission reactions

when they are lowered into or raised out of the nuclear reactor.

They contain _____ to absorb _____.   [3]

c Nuclear power stations produce radioactive waste.
Draw straight lines to match each **type of waste** to its
disposal method.

Type of waste
low level
medium level
high level

Disposal method
Mix it with concrete and store it in stainless steel containers.
Pack it in drums. Dump it in a lined landfill site.
Very difficult to store safely because the radiation damages the container.

[2]

Total [9]

Data and their limitations

1

*It says here that the concentration of nitrogen dioxide in the air in London is more than it is anywhere else in Britain. I don't believe it. In my **opinion**, the idea's a load of rubbish!*

2

*It's not just an idea. My newspaper has lots of **data to justify the statement**. Scientists have measured the nitrogen dioxide concentration in many places.*

3

*OK. But how do we know their data are **reliable**? Do we know **how close to the true value** their measurements are?*

4

Good question. Maybe the measuring instruments were faulty. And even if all the scientists used the same instrument, they could have used it in different ways and got different results. And, of course, the concentration of nitrogen dioxide in any one place changes all the time!

5

So you're saying the data are unreliable?

6

*No. In each place, the scientists took **many measurements**. Then they found the mean. The **mean is the best estimate of the true value**.*

Data and their limitations

1 Each table is designed to collect a set of data about an air pollutant, sulfur dioxide.

Give the letter or letters of the best table or tables in which to collect data to compare

a the concentrations of a pollutant in the four seasons of the year _____

b the concentrations of a pollutant on weekdays and at the

weekend _____

c the concentrations of a pollutant at different times of day _____

d the concentrations of a pollutant in two different cities _____

A

Place: Prague			
Date	Day	Time	Concentration of sulfur dioxide (µg/m³)
3 Jan	Mon	11.00	
6 April	Wed	11.00	
7 July	Thu	11.00	
7 Oct	Fri	11.00	

B

Place: Paris			
Date	Day	Time	Concentration of sulfur dioxide (µg/m³)
3 Jan	Mon	11.00	
5 Jan	Wed	11.00	
6 Jan	Thu	11.00	
9 Jan	Sun	11.00	

C

Place: Brussels			
Date	Day	Time	Concentration of sulfur dioxide (µg/m³)
3 Jan	Mon	11.00	
6 April	Wed	12.00	
7 July	Thu	10.00	
2 Oct	Sun	11.00	

D

Place: Leipzig			
Date	Day	Time	Concentration of sulfur dioxide (µg/m³)
7 July	Thu	08.00	
7 July	Thu	12.00	
7 July	Thu	17.00	
7 July	Thu	22.00	

2 Solve the clues to fill in the grid opposite.
You will need the data in the table to help you fill in some of the words.

Metal	Melting point (°C)	Density (g/cm³)
gold	1063	19.3
lead	327	11.3
iron	1535	7.9
silver	961	10.5
cadmium	321	8.64
zinc	420	7.1

1 Scientists create explanations to account for . . .

2 A student measured the melting point of lead six times. The highest value she obtained was 329 °C, and the lowest value was 324 °C. So the . . . of the readings was 324–329 °C.

3 A measurement that lies well outside the range of the others in a set of repeats is called an . . .

4 Data that are . . . do not vary much when you repeat the measurements.

5 Of the metals in the table, iron has the highest . . .

6 Faulty measuring equipment leads to . . . data.

7 The best estimate of the value of a quantity is the . . . of several repeat measurements.

8 Iron has a . . . density than zinc.

9 The mean of several repeat measurements is the best estimate of the . . . value of a quantity.

10 To get the best estimate of the true value of a quantity, take several . . . measurements and calculate the mean.

11 Of the metals in the table, zinc has the . . . density.

12 All the metals in the table are . . . at room temperature.

13 Scientists use data rather than . . . to justify an explanation.

14 If the mean of a set of readings for one sample is within the range of a set of readings for another sample, there is . . . real difference between the true values of the two samples.

15 A student measured the melting point of a metal in the table several times. She calculated that the mean of her measurements was 959 °C. She concluded that the metal was probably . . .

1 d
2 a
3 t
4
a
5 l
6 i
7 m
8 i
9 t
10 a
11 t
12 i
13 o
14 n
15 s

3 The ideal air temperature of a baby's bedroom is 18 °C. A father is worried that the room is too cold, so he hangs a mercury thermometer on the wall between the curtain and the window. He uses the thermometer to measure the temperature, and gets a reading of 15 °C.

Give three reasons why the measurement may not give you the correct value for the air temperature in the room.

____

____

____

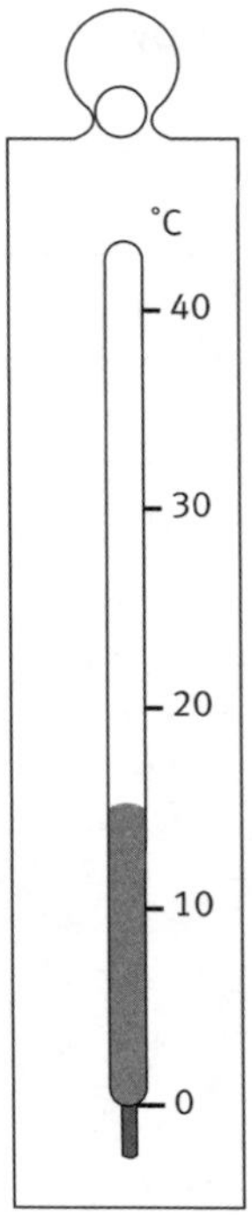
°C
40
30
20
10
0

Data and their limitations

1 Scientists measured the salt content of hamburgers from two restaurants.

They tested six hamburgers from each restaurant.

Their results are in the table.

Sample	Salt content (g)						Range	Average
	1	2	3	4	5	6		
Restaurant A	0.9	1.1	1.0	1.3	0.6	1.2	0.9–1.3	1.1
Restaurant B	1.6	1.4	1.2	1.4	1.3	1.5		

a The scientists work out the range and the average for the samples from restaurant A.
They ignore the value for sample 5.
Suggest why.

_____ [1]

b Work out the range and average for the samples from restaurant B.

Range = _____ to _____ g

Average = _____ g [2]

c The scientists conclude that there is a **real difference** between the salt content of the hamburgers from the two restaurants.

Explain how the data in the table and your answer to part (b) show this.

_____

_____

_____

_____ [1]

Total [4]

2 Five students measured the melting point of tin.
Each student took one measurement.
Here are the results from the five students.

Student	Melting point measurement (°C)
A	233
B	229
C	236
D	230
E	232

a i Calculate the average of the five melting point measurements.

Average = _____ °C   [1]

ii Why is it better to take an average of the melting
point measurements, and not just use the measurement
from one student?
Put a tick in the **one** correct box.

to make it a fair test   ☐

so all students contribute to the result   ☐

to be more confident the result is close to the true value   ☐

because some students are inexperienced in
using thermometers   ☐   [1]

b Work out the range of the melting point measurements.

Range = _____ to _____ °C   [1]

c Another student had a piece of metal. He wanted to
know if it was tin.
He knows that tin always has the same melting point.
He measured the melting point of his piece of metal.
He wrote down this result:

Melting point of my piece of metal = 228 °C.

Was the metal likely to be tin? _____

Give a reason for your answer.

_____   [2]

                                                            Total [5]

3 Ten students measured the concentration of sulfur dioxide
in the air in central Manchester at the same time. Here are
the data they collected.

Student	L	M	N	O	P	Q	R	S	T	U
Concentration of sulfur dioxide ($\mu g/m^3$)	25	27	30	34	22	29	11	26	28	21

a Plot the data on the graph axis below.
One point has been plotted for you.

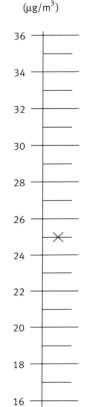

concentration of SO_2
($\mu g/m^3$)

[2]

b i On your graph, circle the outlier for this set of data. [1]

ii Suggest one possible reason for this measurement
being so different from the others.

_____

[1]

c i Ten students measured the concentration of sulfur dioxide in the air in central Birmingham at the same time. Here are the data they collected.

Student	A	B	C	D	E	F	G	H	I	J
Concentration of sulfur dioxide ($\mu g/m^3$)	25	27	30	34	22	29	11	26	28	21

Calculate the average concentration of sulfur dioxide in the air in Birmingham.

Average = _____ $\mu g/m^3$ [1]

ii Work out the range of the sulfur dioxide concentration measurements.

Range = _____ to _____ $\mu g/m^3$ [1]

Total [6]

4 A group of students in the Czech Republic studied data and devised this scientific explanation:

> There is a correlation between being exposed to air pollution and the percentage of sperm with damaged DNA.

a The students collected data on one pollutant, sulfur dioxide. They measured its concentration at the same place every day for six months.
They measurements were different every day.

Why were the measurements different every day?
Put ticks in the boxes next to each possible reason.

Wind direction varied. ☐

A nearby coal-fired power station was running on some days, but not on others. ☐

One student used the measuring instrument incorrectly. ☐

They did not use the same measuring instrument each day. ☐ [2]

b Suggest what other data the students needed to collect to provide evidence for the explanation in the box.

_____

_____ [1]

Total [3]

5 A scientist reads in a journal that shellfish cannot make their shells in sea water below a certain pH.

He knows that when carbon dioxide dissolves in sea water, the pH of the sea water gets lower.

He knows that the concentration of atmospheric carbon dioxide (CO_2) is increasing.

The scientist wants to predict how the changing concentration of atmospheric CO_2 will affect shellfish.

a The scientist studies data on past and present concentrations of atmospheric CO_2. He thinks about the data and predicts future concentrations of atmospheric CO_2.

Which data set is most useful in predicting future concentrations of atmospheric CO_2?

Tick **one** box.

the concentrations of atmospheric CO_2 in 1900 and now ☐

the concentrations of atmospheric CO_2 in 1990, 2000, and now ☐

the amounts of CO_2 released to the atmosphere from the UK in 1990, 2000, and now ☐

the amounts of CO_2 released to the atmosphere from the UK in 1900 and now ☐   [1]

b The scientist assumes that as atmospheric CO_2 concentrations increase, so the concentration of CO_2 that dissolves in sea water will increase.

Suggest what data the scientist can collect to support his assumption.

_____

_____ [2]

c At the end of his work, the scientist makes this prediction:

'By 2050 the pH of sea water will be too low for shellfish to make shells.'

Suggest two reasons why a scientist who studies shellfish in 2050 might find that the shellfish can, in fact, make shells.

_____

_____ [2]

Total [5]

Correlation and cause

China plans to reduce the power station emissions that cause acid rain. It says that acid rain damages its trees, crops, and buildings.

OK. So one of the **outcomes** is that trees are damaged. But surely acid rain is not the only **factor** that affects the amount of damage?

True. There are many other possible factors. Maybe diseases, pests, or even climate change damage China's trees. But there's lots of scientific evidence that acid rain is an important factor.

So changing this factor . . . reducing the amount of acid rain . . . could change the outcome? Fewer trees would be damaged?

Let's hope so!

My company is planning to manufacture bungee-jumping ropes. We need to investigate which material to use. To start with, I want to know which material is strongest.

OK. So you've got a machine that measures the force needed to break ropes made from the different materials?

Yes. And we've got these samples of materials. We'll just put them into the machine.

There'd be a few design flaws if you did that! You need to think carefully about what **factors to control** to make the test fair.

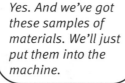

Like all the samples being the same thickness?

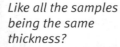

Yes. And they must all be woven into ropes in exactly the same way.

H If you don't control all the factors that might affect the outcome – except the one you're investigating – then your results are meaningless. You will know almost nothing about the relationship between the factor you're investigating and the outcome.

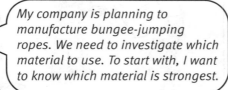

1

Look at this! It says that drinking alcohol gives you mouth cancer! I'm going to lose my tongue and lips – yuk! I'm never going drinking again!

2

Hang on! Not so fast! Read it carefully! It says that that alcohol is a **risk factor** for the disease. Smoking is another one.

3

Ah, it's like we were doing in science. Drinking alcohol is **a factor that increases the chance of the outcome** – getting mouth cancer. But it's not definite that I'll get it.

4

Of course you won't get it. My auntie drank loads. She never got mouth cancer.

5

That's just one case. It tells you almost nothing. You need a much *bigger sample* to provide convincing evidence for or against a correlation.

6

Correlation? What's that?

7

*It's when **an outcome variable increases (or decreases) steadily as an input variable increases**. So as the amount of alcohol drunk by British people increases, so does the number of cases of mouth cancer. You can also say there's a correlation because drinking alcohol **increases a person's chance** of getting mouth cancer.*

8

*It's like this – as the number of hours of sunshine in Blackpool goes up, so does the number of cases of sunburn. That's a **correlation** between a **factor** (hours of sunshine) and an **outcome** (how many people get sunburnt).*

9

Or this survey we did at primary school. We found that taller children are better at adding up than shorter children.

10

*Exactly. They're both correlations. There's probably good evidence that an increase in the hours of sunshine **causes** the number of sunburnt people to increase.*

H *But I doubt if it's like that for your survey. Increasing height is unlikely to cause an increase in children's ability to add up. Probably, they're both caused by some other factor – maybe increasing age.*

1

'Cold noses give you colds,' say Cardiff scientists. Their evidence comes from a simple experiment – getting 90 volunteers to sit with their feet in icy cold water for 20 minutes. Five days later, 29% of them had colds. Only 9% of a control group – who dangled their feet in empty bowls – became poorly.

2

What do you think? Does the factor of having a cold nose increase your chance of the outcome – getting a cold?

*Hmm. Good question. They're **comparing two samples**, which is good. I'd like to know how well the **samples matched**.*

3

You mean it would be a poorly designed study if the average age of one group were much older than that of the people in the other group?

*Exactly. That's a good example. The **sample size** is important too – the bigger the sample, the more confident I'd be in the conclusions.*

4

H *But how can they say 'cold noses give you colds'? The study is all about feet?*

H *The scientists must have thought of a sensible **mechanism to link the factor to the outcome**. If they didn't they wouldn't say that one thing caused the other, even though there's a correlation. Let's read on . . .*

5

H *Here we are. The scientists say that when you get a cold nose, the blood vessels there get smaller. So fewer white blood cells – the ones that fight infection – get to where they're needed.*

H *Clever! So there's good **reason to accept the causal link** that they claim.*

Correlation and cause

1 Circle the letters of the graphs that show a correlation between a factor and an outcome.

A

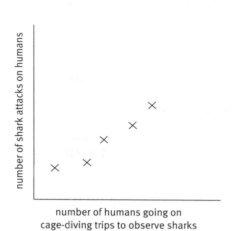

B

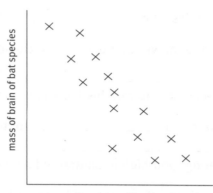

C

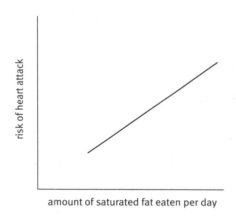

D

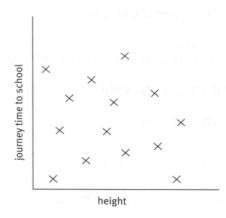

E

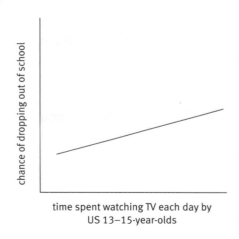

F

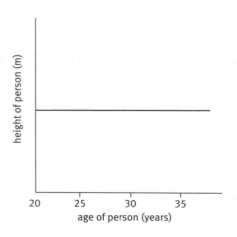

2 Solve the anagrams.

Match each anagram answer to a clue.

Clues
A To investigate whether a factor affects an outcome, we compare samples. The _____ the samples, the more confident we can be in the conclusions.
B If an outcome variable increases as an input variable increases, there is a _____ between the variables.
C Eating food high in saturated fats increases your _____ of having a heart attack.
D When investigating the relationship between a factor and an outcome, it is vital to _____ all the factors that might affect the outcome.
E Scientists usually only believe that a factor causes an outcome if there is a plausible _____ that links the factor to the outcome.
F When comparing samples, the samples must be either _____ or _____.
G Many factors can affect an _____.
H _____ are input variables that may affect an outcome.
I A correlation between a factor and an outcome does not necessarily mean that one _____ the other.

Anagrams
1 for cats
2 cute moo
3 erotic lorna
4 corn lot
5 us aces
6 achenc
7 mad tech
8 erglar
9 can she mim
10 arm nod

Correlation and cause

1 A shampoo company claimed:

> In only 10 days, our anti-break shampoo gives up to 95% less hair breakage than other shampoos.

Scientists at the company tested 10 samples of hair, three times each.

a Identify the factor (independent variable) and the outcome (dependent variable) in the investigation.

Factor: _____

Outcome: _____ [1]

b Another scientist claimed that the investigation was not reliable. Suggest one change the company could make to their investigation so that their results were more reliable.

_____

_____ [1]

c A school student decided to investigate the company's claims. He used this apparatus to test hair strength.

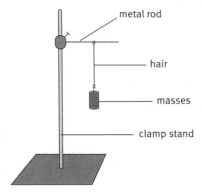

metal rod

hair

masses

clamp stand

i Give two factors the student must control.

_____

_____ [2]

ii Explain why the student must control these factors.

_____ [1]

Total [5]

2 In 2005, scientists studied air quality and staff health in a sample
   of 40 Irish pubs.

   a Write an O next to each of the **three** statements that are outcomes.
      Write an F next to the **one** statement that is a factor that may affect
      the outcomes.

      Since 2004, pub workers have had fewer symptoms
      of smoke irritation, such as watery eyes.   ☐

      In 2004, Ireland banned smoking in all workplaces,
      including pubs.   ☐

      Since 2004, the mean concentration of carbon monoxide
      in pub workers' exhaled breath has decreased by 40%.   ☐

      Since 2004, pub workers' breathing difficulties have
      decreased by 70%.   ☐   [2]

   b i   Use the information in part (a) to complete this sentence:

         There is a correlation between _____

         _____

         and _____

         _____.   [2]

      ii Explain why the correlation you identified in (b)(i) does not
          necessarily mean that the factor caused the outcome.

          _____

          _____

          _____   [1]

                                                                    Total [5]

3 The concentration of ozone in the upper atmosphere has decreased since 1960.

Ozone protects humans from the harmful effects of the Sun's ultraviolet radiation, like getting eye cataracts.

American scientists wanted to know if the 'thinning' ozone layer would lead to more people getting eye cataracts. They studied 2500 people.

Their results showed that **if the concentration of ozone decreases by 20%, the number of people with cataracts is likely to increase by 7%.**

a Tick the boxes next to all the statements that could be true.

The smaller the concentration of ozone in the upper atmosphere, the greater your chance of getting a cataract. ☐

There is a correlation between the concentration of ozone in the upper atmosphere and the number of people with cataracts. ☐

Wearing sunglasses that protect against ultraviolet radiation may reduce your chance of getting a cataract. ☐

If the concentration of ozone in the upper atmosphere increases, the number of people with cataracts is likely to increase. ☐

[2]

b A British scientist wants to find out if there is a correlation between the amount of exposure to the Sun's ultraviolet radiation and the risk of getting a cataract.

She decides to compare two groups of people. All the people in one sample have cataracts; all those in the other sample do not have cataracts.

i Draw a ring around the sample size that will give the most reliable conclusion.

10   100   1000

as large as practically possible

[1]

ii The scientist makes sure that the two samples are matched for age.

Suggest two other factors that she must match.

_____

_____

[2]

Total [5]

4 In December 2005, there was a huge fire at an oil depot in southern England. Some of the chemicals in the smoke were:

carbon monoxide, sulfur dioxide, nitrogen dioxide, solid carbon

a Tick one box next to each statement to show whether the statement is an outcome or a factor that may affect one or more of the outcomes. There are three factors and three outcomes. One has been done for you.

Statement		Outcome	Factor
A	Smoke from the fire rose 3000 m above the ground.		
B	The atmospheric concentration of sulfur dioxide 3000 m over southern England increased.		
C	The atmospheric concentration of nitrogen dioxide near the ground over southern England changed very little.		
D	The smoke was trapped 3000 m above the ground.		✔
E	The atmospheric concentration of carbon monoxide over northern England did not change.		
F	Wind carried the smoke south.		

[2]

b Imagine factor D had been different, and the smoke had not been trapped high above the ground.

In the table below, give the letters of the **two** outcomes that might have been affected.

Then suggest **how** each of these outcomes might have been affected.

Outcome	How the outcome might have been affected

[2]

c   Nitrogen dioxide gas increases the risk of asthma attacks.

Draw straight lines to match each comment with one conclusion.

Comment
1 When the concentration of carbon monoxide increases, there is no change in the number of asthma attacks.
2 More asthma sufferers had asthma attacks in the week after the fire than in a normal week. But not every asthma sufferer had an attack.
3 When the concentration of nitrogen dioxide gas in the air increases, more asthma sufferers have asthma attacks.
4 I know three people who had asthma attacks after the fire.

Conclusion
A There is a correlation between this factor and this outcome.
B These cases do not provide evidence for or against a correlation between the factor and the outcome.
C There is no correlation between this factor and this outcome.
D The factor increased the chance of the outcome but did not always lead to it.

[3]

Total [7]

5 Read the article.

ALCOHOL AND THE RISK OF HEART DISEASE

Research from the 1970s suggested that drinking up to 30 g of alcohol a day led to a 25% reduction in the risk of heart disease. The researchers thought that alcohol increases the amount of 'good' cholesterol in blood, which protects arteries.

In two new studies, scientists have concluded that the reduced risk of heart disease in moderate drinkers could be caused by other factors. For example, 27 of 30 other risk factors for heart disease are higher in non-drinkers than in moderate drinkers.

a i Underline one sentence that states a correlation between a factor and an outcome. [1]

ii Draw a box around one sentence that states that the correlation does not show that the factor causes the outcome. [1]

b i Draw a ring around the sentence that shows that the scientists had thought of a mechanism to link the factor and the outcome. [1]

ii Tick the box next to the statement that best explains why scientists like to think of a plausible mechanism to link a factor and an outcome.

Scientists are more likely to accept that the factor caused the outcome. ☐

Scientists will definitely accept that the factor caused the outcome. ☐

Scientists will definitely not accept that the factor caused the outcome. ☐

Scientists are less likely to accept that the factor caused the outcome. ☐ [1]

Total [4]

Developing explanations

1 It's 1979. Scientists Luis and Walter Alvarez are discussing an exciting find:

This 65-million-year-old rock we found. It's got iridium in it. And the iridium concentration is the same as the iridium concentration in an asteroid.

*That's interesting **data**. Let's think about it.*

2 A few days later:

*I've thought of an **explanation that accounts for our data**. An asteroid collided with Earth 65 million years ago. The collision caused a huge dust cloud that blocked out the sunlight for many years. This killed off all the dinosaurs.*

That's a pretty creative explanation! It links in well to some other data, too: no one has ever found dinosaur fossils in rocks younger than 65 million years old.

3 A few years later:

It says here that scientists have found an asteroid crater in Mexico. And we know that there is iridium in 65-million-year-old rocks all over the world. This extra data has made me even more confident in our explanation!

Why?

I agree about the first part of the explanation – an asteroid colliding with Earth. But I'm less confident about the second part; that the collision made dinosaurs extinct.

Well, there's evidence from other scientists that dinosaurs were already starting to die out before the collision. Also, there haven't been mass extinctions every time an asteroid hit Earth. Maybe it's time to think again...

Developing explanations

1 The statements below describe how a scientist developed an explanation about the origin of life on Earth.

Write the letter of each statement in an appropriate place on the flow chart. Some boxes need more than one letter.

A There are volcanic vents where tectonic plates meet on the sea floor. The vents spew out water, methanol, and other chemicals at about 300 °C.

B Methanol is a simple organic molecule. Normally, methanol molecules break down at 300 °C.

C The volcanic vents contain clay minerals.

D Most scientists believe that life evolved from simple organic (carbon-based) molecules.

E A scientist thought that the clay in volcanic vents might be important. Maybe it could stop methanol breaking down. Perhaps the clay could also help to make bigger organic molecules.

F The scientist thought that, eventually, the clay and organic molecules move out of the volcanic vents. They go to cooler water. Here, life may begin.

G The scientist predicted that, in the lab, clay would protect methanol and help to make bigger organic molecules.

H The scientist made a model undersea volcanic vent in her lab. She observed what happened for six weeks.

I Her prediction was correct.

J Another scientist thinks life is more likely to have started in cooler springs under the ocean. These warm springs have all the ingredients to make organic molecules.

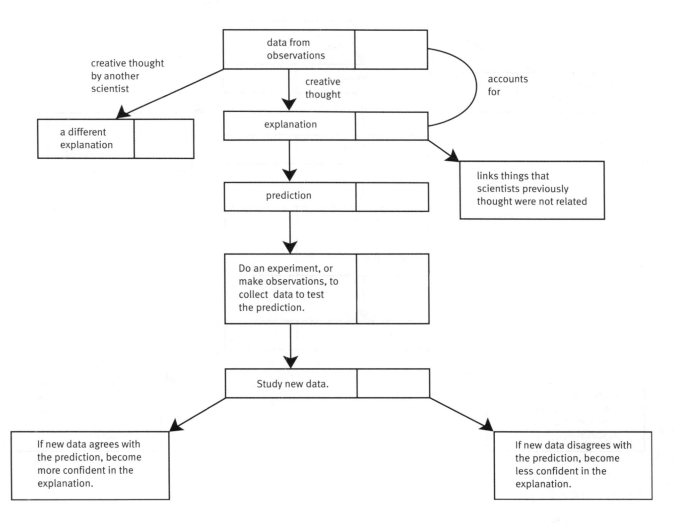

2 The statements below describe how a scientist developed an explanation about mass extinctions.

Write the letter of each statement in an appropriate place on the flow chart. Some boxes need more than one letter.

A Scientists have found 10 enormous lava sheets on the Earth's surface.

B The biggest lava sheet is in Siberia. It is 252 million years old.

C There is lava sheet in India that is 66 million years old.

D No one has found dinosaur fossils in rocks younger than 65 million years old.

E About 251 million years ago 95% of all species became extinct.

F American scientists think that meteorites slammed into the Earth 66 million years ago and 252 million years ago.

G They believe the meteorites pierced the Earth's crust. Lava and carbon dioxide gas rushed out of the holes.

H The scientists think that the extra carbon dioxide gas made the Earth so hot that many species became extinct.

I The scientists predicted that a computer model would show that meteorites had enough energy to pierce the Earth and release enough lava to make the lava sheets.

J The scientists ran the computer model.

K Their prediction was correct.

L Other scientists think that volcanoes erupted in Siberia 252 million years ago. This led to decreased amounts of oxygen in the sea, and caused a mass extinction.

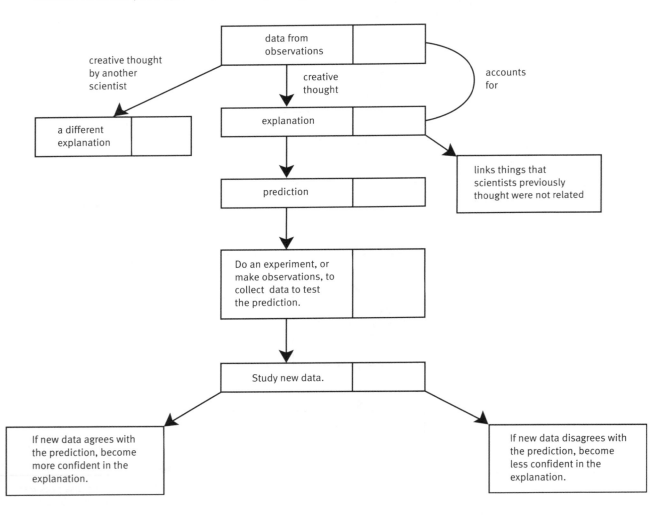

Developing explanations

1 In Kenya, elephants sometimes go onto farmland and ruin maize crops. Some farmers kill these elephants.

Scientists wanted to explain why elephants go onto farmland. They could then predict when elephants were likely to go onto farmland. The scientists could then tell farmers when to make the most effort to guard their crops from elephants.

The scientists studied seven elephants.

They found out where the elephants went and what they ate.

The statements below describe the scientists' work.

A Scientists plan to work with farmers to help them protect their crops from elephants.	
B Satellite tracking showed that six elephants spent all their time in the lowlands.	
C Tail hair analysis from one elephant showed that he ate grass from the lowlands in wet seasons and maize from farmland in dry seasons.	
D In the dry season, there is not enough grass for elephants to eat. Some elephants get the food they need from shrubs and trees. But if there are not enough trees and shrubs, elephants take maize from farmland to eat.	
E Satellite tracking showed that one elephant spent the wet season in the lowlands and the dry season in a forest near farmland.	
F Tail hair analysis of six elephants showed that they ate trees and shrubs in the dry season. In the wet season they ate grass.	
G Fewer elephants will die if scientists find out when they are most likely to eat maize from farms.	

Some of these statements are **data** and one is a possible **explanation**.
Write a **D** next to the **four** statements that are data.
Write an **E** next to the **one** statement that is an explanation.

Total [5]

2 Scientists collected this data:

- ▶ All around the world, rocks from 65 million years ago contain high levels of iridium.
- ▶ Asteroids contain iridium.

From the data, they developed this **explanation**:

> An asteroid hitting the Earth 65 million years ago caused rocks of that age to contain high levels of iridium.

Other scientists used the explanation and extra data to make a prediction:

> **Extra data:** Asteroids and comets were made at the same time.
> **Prediction:** Comet tail dust contains iridium.

The scientists use a spacecraft to collect comet tail dust.
They will examine the dust to find out if it contains iridium.

Tick the boxes next to the **two** statements that are true.

If they find iridium, this will confirm the prediction. ☐

If they find iridium, this will prove the explanation is correct. ☐

If they do not find iridium, this will show that the prediction is definitely wrong. ☐

If they do not find iridium, the prediction may still be correct. ☐

Total [2]

3 Scientists wanted to find out whether smoking increases the risk of developing an eye disease called AMD.
Most AMD sufferers are partially blind.

a The scientists recorded the following statements.

A The greater the number of years a person smokes, and the more they smoke each day, the greater their risk of developing AMD.
B Eye doctors have noticed a build-up of waste substances near the retinas of smokers' eyes.
C 12% of AMD sufferers smoked 20 cigarettes a day for more than 40 years.
D Substances in cigarette smoke may cause damage to cells in the retina of the eye.
E Eye doctors have observed that people with AMD have damaged retinas.

Write the letter **D** next to the **three** statements that are data.

Write the letter **E** next to the **one** statement that is part of an explanation.

Write the letter **C** next to the **one** statement that is a conclusion drawn from data. [3]

b The scientists used their explanation to make this prediction:

> Passive smokers have a greater risk of developing AMD than people who are not exposed to cigarette smoke.

They collected the following data:

▶ Of 100 non-smokers with AMD, 72 were passive smokers.
▶ Of 100 non-smokers without AMD, 66 were passive smokers.

Put a tick in the **two** boxes next to the statements that are true.

The data increase confidence in the explanation. ☐

The data prove the explanation is correct. ☐

The data agree with the prediction. ☐

The data decrease confidence in the explanation. ☐ [2]

Total [5]

4 Nearly 2000 people who lived near Lake Nyos in Cameroon died in 1986.

a Scientists wanted to find out why the people died.
They collected this data:

A Carbon dioxide is soluble in water.

B If carbon dioxide takes the place of air, people die from lack of oxygen.

C There is a volcano below Lake Nyos.

D Carbon dioxide gas is denser than air.

E Sometimes there are small Earth movements near Lake Nyos.

F If you shake a saturated solution of a gas, some of the gas escapes from solution.

G Carbon dioxide has no smell.

H Magma contains dissolved carbon dioxide.

I Carbon dioxide gas is invisible.

The scientists used their data to develop an explanation.
The explanation is in six parts.

Next to each part of the explanation, write one or two letters to show which data each part of the explanation accounts for.

One has been done for you.

Do not write in the shaded boxes.

Part of explanation	Data that this part of the explanation accounts for	
1 Carbon dioxide gas bubbles into the bottom of Lake Nyos.	C	
2 Carbon dioxide dissolves to make a saturated solution.		
3 There was a small Earth movement. This released 80 million cubic metres of carbon dioxide gas from the lake.		
4 Carbon dioxide gas filled the valleys around the lake.		
5 No one detected the carbon dioxide gas, so no one ran away.		
6 1700 people died.		

[8]

b Scientists predict that Lake Nyos will release carbon dioxide again in future. They expect more people to die. They do not know when this will happen.

Suggest one reason why the scientists cannot know when Lake Nyos will next release a large amount of carbon dioxide.

_____

_____ [1]

Total [9]

5 Scientists have studied earthquakes in areas near big, heavy structures made by humans.
They created this explanation:

> Big and heavy structures cause changes in the forces in the ground. These changes may cause earthquakes.

a Study the data in the table.

A	In 1967 there was an earthquake near a huge dam in India. The dam had just been finished.
B	Since 2003, there have been two earthquakes near the world's tallest building, Taipei 101. Taipei 101 was finished in 2003. Before Taipei 101, there were very few earthquakes in the area.
C	In 1967 there was an earthquake under a US mountain. A company had just injected huge amounts of waste into the mountain.
D	The earthquakes under Taipei 101 happened 10 km underground. The building does not affect underground forces at this depth.
E	In 2001 there was an earthquake in the North Sea. Companies have taken many tonnes of oil and gas from the area.

i Give the letters of two pieces of data that the explanation accounts for.

_____ [2]

ii Give the letter of one piece of data that conflicts with the explanation.

_____ [1]

H **b** Governments plan to store carbon dioxide in big underground holes. A scientist uses the explanation in the box to make this prediction:

There will be more earthquakes near carbon dioxide storage holes than there were in these areas before the storage holes were made.

The scientist collects data from now until 2020. Imagine the data shows that there had been five earthquakes near carbon dioxide storage holes. Before the holes were made, no earthquakes had been recorded in these areas.

Put ticks in the boxes that are true.

The data would not prove the explanation is correct. ☐

The data would agree with the prediction. ☐

The data would increase confidence in the explanation. ☐

The data would decrease confidence in the explanation. ☐ [2]

Total [5]

The scientific community

1

New research shows that the brainier male bats are, the smaller their testicles. Scientists studied 334 bat species, and found a correlation between brain size and testicle size: species with small brains have big testicles. The testicles of one bat species account for 8% of male bats' body mass. The scientists reported their findings in a **peer-reviewed scientific journal**.

2

So what do you think of that? That's equivalent to a man's testicles weighing about a stone! More than 6 kg! It can't be right, surely?

Well, the research is about bats – the scientists don't claim to have found out anything about men! But other scientists must have **evaluated** the claim, because the radio report said that the scientific journal is peer-reviewed.

3

Hmmm. It's a very new idea. I'm not sure I believe it.

I know what you mean. **Ideas that have been around longer are certainly more likely to be reliable**. For one thing, there's then been more time for the scientific community to evaluate the claims.

4

And do we know if anyone else has repeated the research?

Not as far as I know. If other scientists did a similar study, and the results were pretty much the same, then there'd be less reason to question their claim.

It would certainly be possible to try to **replicate the research** – the scientists reported exactly what they did very clearly.

THE HISTORY PROGRAMME
Presenters' script

23/08/2012

Presenter 1 (Simon)

Welcome to *The History Programme*. This year, 2012, is the hundredth anniversary of Wegener's theory of continental drift. We all now recognize the importance of his ideas. But a century ago scientists laughed at him.

Presenter 2 (Janet)

Yes, that's right. Wegener explained that the east coast of South America was once joined to Africa's west coast. The two continents had been slowly moving apart ever since. Wegener had lots of data to support his explanation: the shapes, rock types, fossils, and mountain ranges of the two continents matched up closely.

Presenter 1

So why did other scientists disagree with Wegener?

Presenter 2

Well, of course you can't simply deduce explanations from data. So it's quite reasonable for different scientists to come to different conclusions, even if they agree about some of the evidence. But there's more to it than that.

Presenter 1

Tell me more.

Presenter 2

It seems that other scientists simply couldn't imagine how massive continents could move across the planet. It was an **idea outside their experience**. Also, they didn't much respect Wegener – he was never regarded as a **member of the community of geologists**.

Presenter 1

And I suppose scientists don't give up their 'tried and tested' explanations easily?

Presenter 2

Exactly. Scientists often feel that it's safer to stick with **ideas that have served them well in the past**. Of course, new data that conflict with an explanation make scientists stop and think – but it could be that the data are incorrect, not the explanation! Generally, scientists only abandon an established explanation when there are really good reasons to do so, like someone suggesting a better one.

The scientific community

1 The stages below describe one way a scientific discovery is made and then accepted by other scientists.

They are in the wrong order.

A The scientist tells other scientists about the investigation results, either at a conference or in a scientific journal.

B Other scientists repeat the investigations.

C If their results are similar, the other scientists accept that the new idea is correct.

D Other scientists ask questions and evaluate the scientist's claims.

E A scientist makes an unexpected observation.

F The scientist does further investigations.

Fill in the boxes to show the right order. The first one has been done for you.

E					

H 2 Write a C next to reasons why two scientists may come to different conclusions about the same data.

Write an X next to reasons for scientists not abandoning an explanation even when new data do not seem to support the explanation.

a The scientists are interested in different areas of science. ☐

b The data may be incorrect. ☐

c The new explanation may run into problems. ☐

d Different organizations paid for each scientist's research. ☐

e It is safer to stick with ideas that have served well in the past. ☐

3 Below are eight answers. Make up one question for each answer.

Peer review

_____

Scientific conference

_____

Other scientists get similar results.

_____

The explanation has stood the test of time.

_____

Scientific journal

_____

A tobacco company employs one scientist; a cancer charity employs the other scientist.

_____

New data may be inaccurate.

_____

Different sponsors have paid for the research.

_____

4 Read the information in the box.

> It's not only fatty foods and smoking that are risk factors for heart disease! New research shows that decaffeinated coffee may also be bad for your heart. American scientists studied 187 people for three months. They found increased levels of harmful cholesterol in the blood of people who drank decaffeinated coffee. The researchers presented their findings to other scientists at the American Heart Association's conference.

Make up a dialogue on page 137 to get across six important points about the scientific community.

Use the information in the box above and the phrases below.

- ▶ scientific journal or conference
- ▶ peer review
- ▶ evaluated by other scientists
- ▶ repeatable results
- ▶ different conclusions about the same data
- ▶ scientists not wanting to give up an explanation that has stood the test of time

The scientific community

1 Read the information.

> ▶ British scientists studied the effects of combinations of food additives on nerve cells.
> ▶ They used nerve cells from mice.
> ▶ They found that some combinations of food additives made the nerve cells stop growing.
> ▶ They found that the effects of combinations of additives were greater than the effects of single additives.
> ▶ They published their findings in a scientific journal.
> ▶ In the article, the scientists mentioned that food additives have been linked to behaviour problems in children.

  a Why do scientists report their findings in scientific journals?
     Tick the boxes next to the best answers.

      to make sure their data is correct   ☐

      so that other scientists can repeat the research   ☐

      to make sure their research is reliable   ☐

      so that other scientists can evaluate the research   ☐   **[2]**

  b **i** A scientist reads the journal article carefully.
       She wants to repeat the research.
       Give one thing she needs to find out from the article before she can begin her experiment. Do not include things mentioned in the box above.

       _____

       _____   **[1]**

H   **ii** Why do scientists believe it is important to repeat experiments done by others?

       _____

       _____   **[1]**

c Other organizations commented on the research described in the box.

Draw lines to match each comment to an organization that might have made the comment.

Organization
1 Food Standards Agency
2 Representative from a sweeteners information service
3 Food and drink manufacturers

Comment
A The research did not provide meaningful information. The mice were given undigested aspartame.
B The European Union's science committee says that all the additives the scientists used are safe.
C We need more details about the research before we can assess its value. We are also funding research about the effects of combinations of food additives.

[2]

Total [6]

2 Read the article.

> ## Vitamin D cuts cancer risk
>
> US scientists looked in medical journals to find reports on research linking vitamin D levels to cancer rates. They found 63 reports from 1966 to 2004. Each study showed that people who do not have enough vitamin D are more likely to develop certain cancers than people who do have enough vitamin D.
>
> Sunlight helps our bodies to make vitamin D. A glass of milk contains 100 units of vitamin D.
>
> The scientists calculated that taking a vitamin D supplement of 1000 units each day may cut by 50% the risk of getting some cancers. They warn that more than 2000 units of vitamin D each day can damage the liver, bones, and kidneys.

a **i** Underline the sentence that mentions one way in which scientists report their findings to other scientists.

    ii Circle the two sentences that show that many scientists collected similar evidence and came to similar conclusions.

    iii Draw a box around the paragraph that advises on the daily dose of vitamin D that cuts cancer risks.

[3]

b The article claims that a daily vitamin D supplement cuts the risk of getting some cancers.

How strong is the evidence supporting this claim?

_____

Give one reason for your answer.

_____

_____

[1]

c In 2015, a scientist collects new evidence about the link between vitamin D and cancer risks. She claims that her evidence shows that vitamin D intake and cancer risks are not related. This contradicts all previous evidence.

Give one reason why other scientists might not want to accept this claim.

_____

_____

[1]

Total [5]

3 Read the information about the work of Charles Darwin in the1860s.

> Darwin collected evidence by making these observations:
>
> ▶ The individuals of a species are slightly different from each other.
> ▶ There are always more members of a species than can survive.
>
> From his observations, Darwin developed his **theory of natural selection**:
>
> ▶ 'Any variation that helps an individual to survive is more likely to be inherited by its offspring'.
>
> Darwin's theory helps to explain how and why evolution happens.

Below are some statements from other scientists about Darwin's theory. Some statements are from 1870; others are from 2007.

A I don't know why there is variation within a species.

B There has not yet been time for other scientists to evaluate Darwin's ideas.

C DNA evidence explains how species are related to each other.

D I disagree with Darwin's theory. God created everything on Earth in seven days.

E We know that the Earth is about 4 thousand million years old.

F Darwin's observations are fine. But I don't know how living things pass on variations to their offspring.

G We often find new fossils, and we can date them accurately now.

H Darwin's theory has worked well for more than a hundred years.

I There is not enough fossil evidence to support Darwin's ideas.

J The Earth has not existed long enough for evolution to have happened.

K The data may be incorrect.

a Give the letters of four statements that show why many scientists in 1870 did not support Darwin's theory of natural selection.

_____

_____

_____

_____

[4]

b Give the letters of three statements that mention **evidence** that might make a 2007 scientist more likely to accept Darwin's theory than an 1870 scientist.

_____

_____

_____

[3]

Ⓗ c In 2020, a scientist collects data that seems to contradict Darwin's theory. He proposes a new explanation for evolution. Most scientists do not immediately accept the new explanation.

Give the letters of two statements that are reasons for scientists not accepting the new explanation.

_____

_____

[2]

Total [9]

Risk

Don't buy that drink – it contains E211, that's sodium benzoate. We don't want to risk the kids' asthma and eczema getting worse.

But **nothing is risk free**. Many scientific advances introduce new risks. Even if the kids drink only water, there's a risk of harm. A poisonous chemical could contaminate it.

True, but it's so unlikely. I read about a Food Standards Agency report. They found that E211 can make the symptoms of asthma and eczema worse in children who already have them. Our kids' eczema and asthma are bad enough already. Let's minimize the risk of them getting worse.

Why are you lying out there in the sun at midday? Where's your sunscreen? It's dangerous to sunbathe – ultraviolet radiation from the sun hugely increases your risk of getting skin cancer. Each year, nearly 6000 British people get melanoma skin cancer.

Yes, but there are **benefits** too. Sunlight helps you make vitamin D. You need this to strengthen your bones and muscles, and to boost your immune system. Anyway, I feel more confident when I've got a tan. And it's lovely and warm out here...

OK. I guess it's up to you. But I wish you wouldn't take the **risk**!

H

1

The government is planning to build new nuclear power stations. A spokesperson says that they will safeguard future electricity supplies without contributing to global warming.

2

Well, they'd better not build one near here. I'm moving if they do.

What's your problem? The **chance** of a nuclear power station exploding is tiny.

3

But you've got to think of the **consequences** if it did explode. Two and a half thousand people died after the 1986 accident at Chernobyl.

And don't forget the risk of radioactive materials leaking out.

4

You're getting confused here. The actual measured risk of a leak is very low. I reckon you **think the risk is much higher than it really is** because ionizing radiation – and its effects – are **invisible**.

5

We've not even mentioned nuclear waste! We can't possibly know about the effects of this over the next thousand years.

Exactly. The government must not build them. It's simple – we just apply the **precautionary principle!** Better safe than sorry.

6

Have you forgotten the **ALARA principle**? The government needs to decide what level of risk is acceptable. Then it has to spend enough money to reduce the risk to this level. That's it – the risk will be As Low As is Reasonably Achievable!

Risk

1 Use the clues to fill in the grid.

1 Everything we do carries a risk of accident or h . . .

2 We can assess the size of a risk by measuring the c . . . of it happening in a large sample over a certain time.

3 Mobile phones are an example of a technological a . . . that brings with it new risks.

4 Radioactive materials emit i . . . radiation all the time.

5 The chance of a nuclear power station exploding is small. The c . . . of this happening would be devastating.

6 To apply the . . . principle, people must decide what level of risk is acceptable.

7 Some people think that the effects of a nuclear power station explosion are so terrible that the government should adopt the . . . principle and not build them at all.

8 Sometimes people think the size of a risk is bigger than it really is. Their perception of the size of the risk is greater than the . . . risk.

9 Nuclear power stations emit less carbon dioxide than coal-fired power stations. Some people think that this b . . . is worth the risk of building nuclear power stations.

10 Many people think that the size of the risk of flying in an aeroplane is greater than it really is. They p . . . that flying is risky because they don't fly very often.

11 It is impossible to reduce risk to zero. So people must decide what level or risk is a . . .

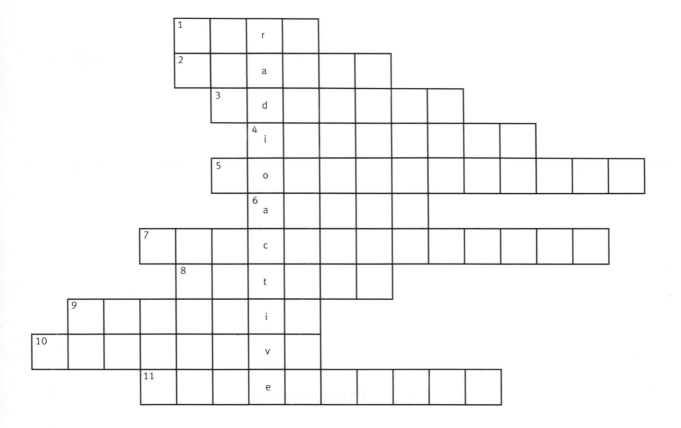

2 Draw a line to link two words on the circle.
Write a sentence on the line saying how the two words
are connected.
Repeat for as many pairs as you can.

risk

safe

benefits

chance

consequences

balance

unfamiliar

scientific advances

precautionary principle

actual risk

perceived risk

ALARA principle

Risk

1 Research shows that using sunbeds increases the risk of developing skin cancer. Ultraviolet radiation from sunbeds can cause eye cancer.

Some people who use sunbeds suffer from dry, bumpy, or itchy skin.

Many young people use sunbeds in tanning shops regularly.

a One tanning shop displays this notice.

Using our sunbeds

▸ Lock the door. ☐

▸ Wear goggles. ☐

▸ Use the sunbed for a maximum of 15 minutes. ☐

▸ Wipe the sunbed clean after use. ☐

▸ Return your towel to reception. ☐
▸ Do not use the sunbed if you have lots of moles or freckles. ☐

Some of the information describes ways of reducing the risks of using sunbeds.

Write **R** in **three** boxes next to sentences that describe ways of reducing the risks. [3]

b i Suggest two reasons why many young people are willing to accept the risks of using sunbeds.

Reason 1 _____

_____

Reason 2 _____

_____ [2]

H

ii Three people made these comments about the risks of using sunbeds.

   A I reckon the risk of getting skin cancer from sunbeds is tiny. My mum uses them all the time. She wouldn't if they were dangerous!

   B I read an article in a medical journal. It says that the risk of developing skin cancer increases by 20% for every ten years you use sunbeds.

   C Sunbeds are really dangerous. Your skin can go red and itchy when you use them. You can get eye cancer from them really quickly, too.

   Give the letter of **one** comment that identifies an

   actual risk of using sunbeds. _____

   Give the letter of **one** comment that identifies a

   perceived risk of using sunbeds._____   [2]

c A Member of Parliament (MP) wants to ban under-18s from using sunbeds.

He does not want to stop adults from using them.

Use ideas about risk, benefit, and balance to discuss possible reasons for the MP wanting to ban under-18s, but not adults, from using sunbeds.

One mark is for a clear and ordered answer.

_____

_____

_____

_____

_____

_____   [3 + 1]

Total [11]

2 The Food Standards Agency (FSA) has found out that *Campylobacter* is the most common cause of food poisoning in Britain. Sufferers have severe diarrhoea and abdominal cramps.

The FSA did some research about *Campylobacter* in chickens. They bought chickens from shops. They found *Campylobacter* bacteria in 56% of the fresh chickens and 31% of the frozen chickens.

a The table lists ways of reducing the risk of becoming infected with *Campylobacter*.

Write the letter **I** next to actions that individual people can take.

Write the letter **G** next to actions that Governments can take.

A Cook chicken thoroughly.	
B Do not eat chicken.	
C Put advertisements on the radio to persuade people to cook chicken thoroughly.	
D Tell chicken farmers to apply very strict hygiene measures on their farms.	
E Wash your hands thoroughly after handling raw chicken.	
F Buy frozen chicken instead of fresh chicken.	

[2]

b Suggest two reasons why many people eat chicken, even when they know that this increases their risk of becoming infected with *Campylobacter*.

_____

_____

[2]

H

c The quotes show how four people react to the risk of becoming infected with *Campylobacter*.

Angus

I have heard that there is a risk of becoming infected with Campylobacter from eating chicken. So I will always make sure that the chicken I eat is properly cooked.

Brendan

I read in the newspaper that some Campylobacter can lead to severe long-term illness in a few people. I don't want to increase my risk of becoming very ill. So I will never eat chicken again.

Callum

I know that eating chicken carries a risk of becoming infected with Campylobacter. So I will take sensible precautions when I handle raw chicken.

Douglas

Only 1% of British people are infected with Campylobacter each year. So I don't believe the risk is worth worrying about.

The quotes show that one person is applying the precautionary principle.
Give his name.

_____

[1]

d A government has decided to apply the ALARA principle to the risk of becoming infected with *Campylobacter*. Write a sentence that the public can understand to state the government's policy on infection by *Campylobacter*.

_____

_____

_____

[2]

Total [5]

3 Read the article.

Food for the brain

A scientific report links poor diet to problems of behaviour and mood. The report says that people who do not eat enough vitamins, minerals, and essential fats are at increased risk of attention deficit disorder, depression, and schizophrenia.

The report advises eating plenty of fruit, vegetables, fish, and seeds to prevent mental health problems. It also suggests not eating 'junk food', like burgers and chips.

The scientists want the government to make sure that school meals include 'brain healthy' food to help prevent young people getting mental health problems.

a i Draw a ring round the paragraph that identifies some risks of a poor diet.

ii Draw a box around the paragraph that suggests ways of reducing the risks of getting mental health problems. [2]

b The article asks the government to take action.

Use ideas about risk, costs, and benefits to write a paragraph arguing **for** or **against** this action.

One mark is for quality of written communication.

_____

_____

_____

_____

_____

_____ [3 + 1]

Total [6]

4 Human activities cause an increase in greenhouse gases, such as carbon dioxide. Increasing amounts of greenhouse gases lead to global warming. Global warming causes rising sea levels and flooding.

The graphs show how greenhouse gas emissions from the USA and the European Union changed between 1990 and 2000.

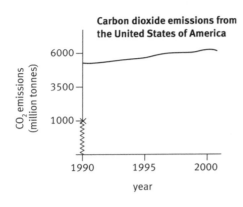

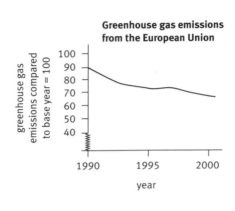

a Describe the trend for the European Union.

_____

_____

[1]

H **b** Complete the sentences.

Use words from this list.

benefits	chances	consequences	precautions

If all countries have a similar trend to the USA, the _____

that the Earth's temperature will continue to increase are high. The

_____ of this for people who live on low-lying islands

could be devastating.

[2]

c Some low-lying islands might disappear completely because of flooding caused by global warming. Are people who live on these islands more likely to want everyone in the world to adopt the **ALARA principle** or the **precautionary principle** towards reducing the risk of global warming? Give a reason for your choice.

_____

_____

_____

_____

[2]

Total [5]

Making decisions about science and technology

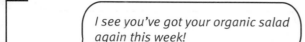

1

> I see you've got your organic salad again this week!

> Yes. It's delicious. And the farmer uses no synthetic fertilizers or pesticides. So there's no risk of me eating chemicals – like pesticides – that may be harmful.

2

> OK. So people like you – who can afford organic food – benefit. And farmers sell organic food at higher prices than non-organic food. But what about people who need cheaper food?

3

> Exactly. Farmers who use synthetic fertilizers and pesticides get a much higher yield from one square metre of land. So their food is usually cheaper. And there are no bugs in their carrots!

> True. There are always **benefits and costs**. We need to weigh them up. Intensive agriculture can sometimes damage the soil and pollute water.

4

> And organic farming is sustainable?

> I reckon so. **Sustainable development** just means being able to meet people's needs now, and at the same time leaving enough for the future.

> OK. But how do you know your salad really is organic?

> Easy! The Soil Association has checked up on it – they have very strict regulations.

Debating Society: 18 March

Is it ethically acceptable to clone human embryos to produce stem cells to treat illnesses?

1

I will start with a short explanation. Embryonic stem cells are cells that are not yet specialized; they can develop into any type of cell. To obtain embryonic stem cells, scientists create embryos in the lab. They harvest stem cells from the embryos. Then they destroy what's left of the embryos.

2

*I believe that that, ethically, the process is totally unacceptable. Life begins at conception. It is wrong and **unnatural** to create and destroy embryos in the lab. I believe this even though I know that stem cell therapy may greatly benefit some people.*

3

As we have just heard, stem cell therapy has many likely benefits. Stem cells could provide tissue for an organ transplant to save someone's life. They could treat heart disease or diabetes.

4

So what do you think, ethically?

5

*I believe that the right decision is the one that leads to the **best outcome for the majority of people involved**. So if many people will benefit greatly, it is right to go ahead with researching stem cell treatments.*

6

Even if some people – or embryos – suffer?

Yes.

I'm not so sure.

Making decisions about science and technology

1   Tick the boxes next to the questions that a scientist or technologist could try to answer.

   A   Should Britain build more nuclear power stations, or concentrate on renewable energy resources instead?   ☐

   B   Could nuclear power meet all Britain's current energy needs?   ☐

   C   What techniques for cloning stem cells are successful?   ☐

   D   Should the MMR vaccination be compulsory?   ☐

   E   Should I buy organic vegetables or those that have been intensively grown?   ☐

   F   How much plastic waste could be recycled in Britain?   ☐

   G   If an AIDS vaccine were available, who should be given it?   ☐

   H   What percentage of children who get the MMR vaccine get ill with measles?   ☐

   I   Is it ethically right to clone stem cells to treat disease?   ☐

   J   Should Britain recycle more of its plastic waste?   ☐

   K   Is it possible to develop an AIDS vaccine?   ☐

   L   How much wheat can be grown in this field using intensive farming techniques?   ☐

2   Match each statement to the idea it shows.

Statement	Idea it shows
1 All 11-year-olds should have a flu vaccine. This would protect most of the population from flu, even elderly people who weren't vaccinated themselves.	A It is unfair for some people to have to take the risk so that everyone will benefit.
2 No one should be vaccinated against flu. Vaccinations interfere with nature.	B This decision is the one that benefits the most people.
3 Occasionally, the flu vaccine has side effects. Why should it be only 11-year-olds who are put at risk, when nearly the whole population would benefit?	C Things that are unnatural are never right.

3 Find 12 words from this Ideas about science section in the wordsearch. Then write a crossword clue for each word.

S	S	N	O	I	T	S	E	U	Q	T
U	N	A	R	T	B	R	I	S	K	E
S	O	B	N	A	E	H	E	R	I	C
T	I	E	E	C	N	Q	S	E	N	H
A	T	V	Y	D	E	U	I	G	S	N
I	A	O	A	N	F	T	E	L	E	I
N	L	L	A	C	I	H	T	E	U	C
A	U	N	N	A	T	U	R	A	L	A
B	G	L	C	O	S	T	S	X	A	L
L	E	L	B	I	S	A	E	F	V	L
E	R	I	M	A	J	O	R	I	T	Y

Word	Clue

4 Read the information in the box about values and ethics in science. Then make notes about values and ethics in science in the table.

▶ Write a title in the top row.
▶ Write the two or three most important points in the next row down.
▶ Write other, detailed, information in the lowest three rows.

There are many questions that cannot be addressed using a scientific approach. For example, a scientist can find out how to get stem cells from embryos, and how to use the stem cells to treat diseases. But people have different views about whether it is ethically right to actually use these techniques. So it is up to others – not just scientists – to answer the question *'is it ethically acceptable to use embryonic stem cells to treat disease?'*

People use different sorts of arguments when they discuss ethical issues. One argument is that the right decision is the one that gives the best outcome for most people. Another argument is that some actions are unnatural or wrong; it is never right to take these actions. Thirdly, some people think that it is unfair for one person to benefit from something when others have taken a risk.

Title:	
Most important points:	
Other information:	

Making decisions about science and technology

1 **a** An electricity company needs to decide whether to build a new nuclear power station or a coal-fired power station.

The company asks people from five organizations for their opinions.

Freya Electricity from coal and nuclear power stations costs about the same.

Grace Coal-fired power stations lead to acid rain that damages our beautiful buildings.

Hanif Nuclear power stations produce much less carbon dioxide gas than coal-fired power stations.

Ian There is no totally safe way of getting rid of nuclear waste.

Jasmine Nuclear power has its advantages, but an accident at a nuclear power station could kill thousands.

Write the names of the people who would probably prefer the company to build a new coal-fired power station, not a nuclear power station.

_____ [2]

b Old nuclear power stations must be taken down.
This process is called decommissioning.

Tick the boxes next to the jobs that are likely to be part of the work of the Nuclear Decommissioning Authority.

making sure that used fuel rods are disposed of safely ☐

deciding whether to build new nuclear or gas-fired power stations ☐

regularly checking storage sites where low-level and intermediate-level radioactive waste are stored ☐

deciding which countries to buy nuclear fuel from ☐ [2]

Total [4]

2 Pre-implantation genetic screening (PGS) is a technique to choose the best embryo to implant in a woman who is having fertility treatment.

The technique involves removing one cell from an eight-cell embryo.
Scientists then test this cell for abnormal chromosomes.
Abnormal chromosomes may cause conditions like Down's syndrome.

a Five people were asked for their opinions about PGS.
Some people support the technique; others think it
should not be allowed.

Kirsty

Linda

PGS is not natural.

It is not fair for parents to benefit by putting their embryos at risk – embryos are people too! The parents don't risk much themselves.

It's better to stop babies with genetic defects being born; it's expensive for society to look after them as they grow up.

Marcus

Nikki

People with conditions like Down's syndrome have a lot to offer. It is wrong to prevent them being born.

People with chromosome abnormalities can have very difficult lives. It's better for them – and their parents – if they are not born in the first place.

Oliver

Write the names of the people in the correct columns in the table.

People who *agree* with pre-implantation genetic screening	People who *disagree* with pre-implantation genetic screening

[3]

b Write **T** next to the questions that address **technical issues** about PGS.

Write **V** next to the questions that address issues of **values** of PGS.

Is PGS ethically acceptable? ☐

Does the technique damage embryos? ☐

Is it natural to choose which embryo to implant in a woman's uterus? ☐

Do embryos that have been tested grow properly when they are implanted into a woman's uterus? ☐

Is PGS necessary – maybe embryos can fix their own genetic defects? ☐

Is it right to destroy embryos that are not implanted? ☐

[3]

c Suggest two reasons why PGS is offered to parents in the UK but is not offered to parents in some other countries.

_____

_____

_____

_____

[2]

Total [8]

3 Sustainable development is planning how to meet people's needs now without damaging the environment for people in future.

a Tick the boxes next to the statements that show some of the ways in which organic farming is sustainable.

Organic farming does not use synthetic chemical fertilizers that can damage the soil's structure if used carelessly. ☐

Some people think that organic food tastes better. ☐

Organic food is more expensive than food that is not produced on organic farms. ☐

Organic farmers control pests with natural predators, not synthetic pesticides. ☐

Organic food does not have pesticide residues on it. ☐

[2]

b A power station in Bristol burns hospital waste to generate electricity.

i Suggest one way in which generating electricity from hospital waste is more sustainable than simply burning hospital waste.

_____

_____

[1]

ii It is technically possible to use waste from all British hospitals to generate electricity.

However, waste from only a few hospitals is used in this way.

Suggest two reasons for this.

_____

_____

_____

_____

[2]

Total [5]

H 4 a Malaria kills millions of people every year.
There is no vaccine against malaria.
Below are some reasons for this.

Malaria parasites become resistant to new drugs and vaccines very quickly. ☐

Malaria mainly affects people living in economically poor countries. ☐

There are several different malaria parasites. Each needs a different vaccine. ☐

Drug companies make more money from selling drugs for heart disease than they could make from a malaria vaccine. ☐

Governments of economically rich countries are unwilling to spend money on scientific research to develop a malaria vaccine. ☐

Write **T** next to the statements that give **technical** reasons for there being no malaria vaccine.

Write **V** next to the statements that give **values-related** reasons for there being no malaria vaccine. [3]

b All British babies are offered the MMR vaccine against measles, mumps, and rubella.
Some parents decide not to have their babies vaccinated.

i Give one reason why some people think that parents should be free to choose whether or not their baby gets the MMR vaccine.

_____

_____

_____ [1]

ii Give one reason why some people think that the MMR vaccine should be compulsory.

_____

_____

_____ [1]

Total [5]

Case study

The case study is your chance to find out more about a science-related issue that interests **you**. It's worth 20% of your total mark.

Choosing a topic

You need to find a science topic that is controversial – one that people have different opinions about. To get ideas, look in newspapers and magazines, or pick things up from television, radio, or the Internet.

It's best to choose from one of the topic types in the table.

Topic type	Examples	What to focus on
evaluating a claim where scientific knowledge is uncertain	▶ Is there life elsewhere in the Universe? ▶ Does using mobile phones increase risk of brain damage?	▶ relationships between data and explanations ▶ the quality of research behind different scientists' claims
helping to make a decision on a science-related issue	▶ Should a shopping street be pedestrianized to reduce air pollution? ▶ Should pre-implantation genetic screening be available free to anyone who wants to have a baby? ▶ Should Britain build new nuclear power stations?	▶ personal choice and values ▶ balancing risks and benefits of possible action
personal or social choices	▶ Should my child receive the MMR vaccine? ▶ Should I recycle all the plastic I use?	▶ personal and ethical issues ▶ using science to evaluate these issues

For the title, make up a question that you can answer by balancing evidence and opinions.

Selecting and using information (4 marks)

▶ Choose sources of information that are
  – varied, for example books, leaflets, newspaper articles, and websites
  – reliable: research reports from a university website may be more reliable than an individual's blog; journalists have their own opinions and don't always give balanced views; an organization that pays for a piece of science research may influence the research findings
  – relevant: if something is not relevant, throw it out!
▶ At the end of your report, include references to every source. Make it easy for someone else to find the information you have used (and check up on you!).
▶ Throughout your report, give the exact source of every quotation and opinion.

The science of the case (8 marks)

▶ Check what the scientific knowledge you need in order to understand the issues in your study. You should be able to find most of it either in your Textbook or in another source written at a similar level.

▶ Consider how well (or badly!) each opinion that you describe is supported by a science explanation.

▶ Look carefully at the quality of scientific evidence in each source to judge whether its claims are reliable.

In your report, you need to show that you've done all this – doing it but not writing about it counts for nothing! So include plenty of detail, and make sure that you link every claim or opinion to relevant scientific evidence.

If the science is too difficult, it is best to choose another topic.

Your conclusions (8 marks)

This is another chance to show how well you've understood Ideas about science, particularly: data and their limitations; risk; making decisions.

In your conclusion

▶ Compare opposing evidence and views.
  – Report and evaluate arguments 'for' and 'against'
  – Compare these arguments carefully and critically
▶ Give conclusions and recommendations.
  – Suggest two or three different conclusions to show you realize that evidence can be interpreted in different ways.

Presenting your report (4 marks)

First of all, decide on an 'audience' for your report – this could be Year 9 students, a Member of Parliament, or any other individual or group.

Depending on the resources available, you might produce a formal written report, a newspaper article, a poster, or a PowerPoint presentation. Think about what is most appropriate for your audience and what you want to tell them. Whatever method you use, make sure that it looks attractive!

Then work on the following:

▶ The structure and organization of your report:
  – Put everything in a logical order, with plenty of subheadings.
  – Include page numbers and a contents list.
▶ Visual communication:
  – Include pictures, diagrams, charts, or tables to help your audience understand ideas and information.
▶ Spelling, punctuation and grammar
  – Be concise – don't waste words!
  – Use relevant scientific words.
  – Check your spelling, punctuation, and grammar very carefully.

Data analysis

This is your chance to have a go at interpreting and analysing real data. It's worth 13.3% of your total mark.

Getting started

Start by doing a practical activity to collect the data you need. You can do this on your own or in a small group. Once you've collected some data yourself, you may be able to get the rest from other students, a teacher demonstration, or other sources.

Interpreting data

Follow the advice below, and you should get a high mark!

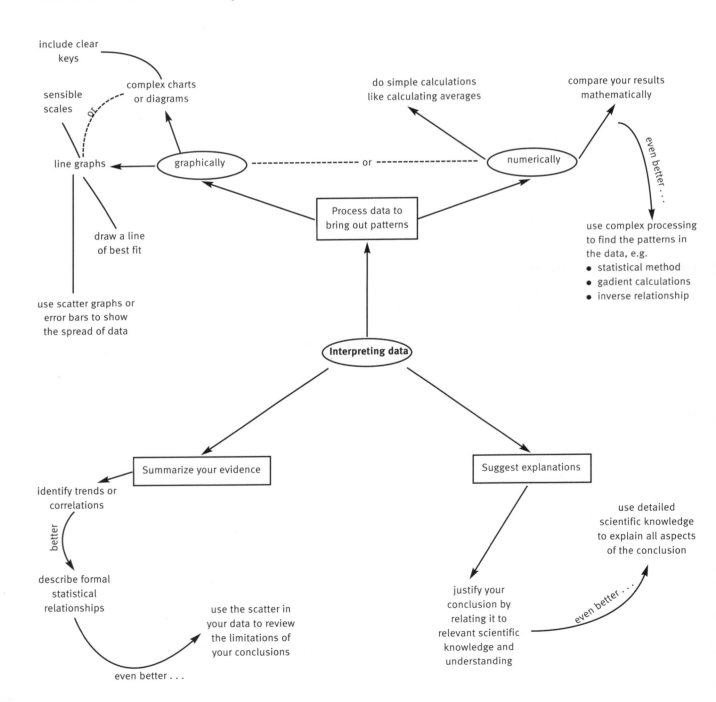

Evaluating data

Follow the advice below to achieve the very best you can!

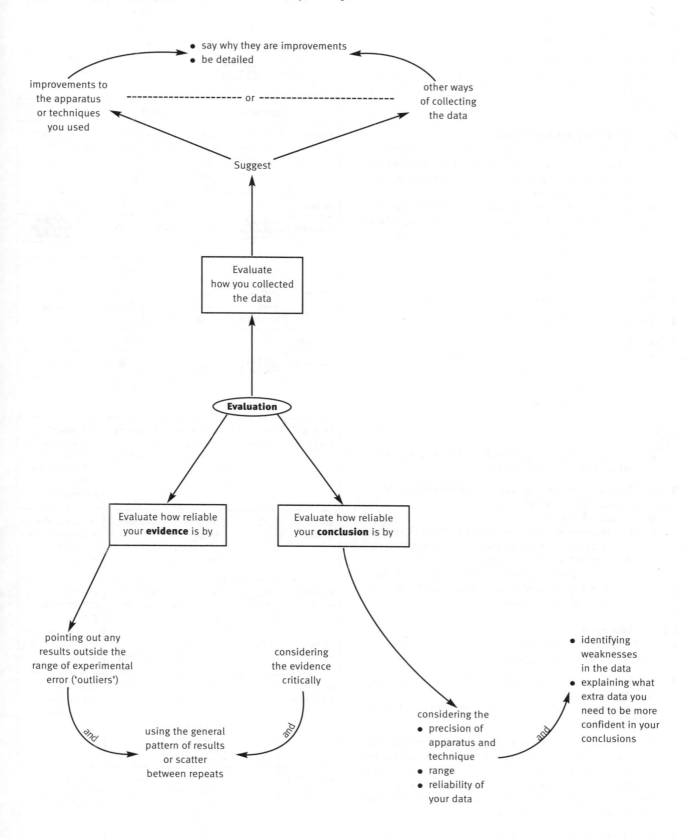

• say why they are improvements
• be detailed

improvements to
the apparatus
or techniques
you used

———————————— or ————————————

other ways
of collecting
the data

Suggest

Evaluate
how you collected
the data

Evaluation

Evaluate how reliable
your **evidence** is by

Evaluate how reliable
your **conclusion** is by

pointing out any
results outside the
range of experimental
error ('outliers')

considering
the evidence
critically

and

using the general
pattern of results
or scatter
between repeats

and

considering the
• precision of
  apparatus and
  technique
• range
• reliability of
  your data

and

• identifying
  weaknesses
  in the data
• explaining what
  extra data you
  need to be more
  confident in your
  conclusions

Answers to questions

B1 Workout

1 **a** Cell **b** Genes **c** Nucleus
   d Chromosome **e** DNA

2 **a** T **b** T **c** F **d** F
   e F **f** T **g** F **h** T

3 Unspecialized, asexual, clones, environments

4 1J, 2N, 3A, 4D, 5G, 6H, 7I or B or F or L, 8B or F or L or I, 9E or P, 10M, 11B or F or L, 12C, 13O, 14B or F or L, 15P or E, 16K

B1 GCSE-style questions

1 **a** They have the same combination of alleles; They both developed from one egg that was fertilized by one sperm; They both started growing from one embryo. The cells of the embryo separated.
   b A person's characteristics are affected by both genes and the environment; They received alleles from both parents; The twins and their mother have different combinations of alleles.
   c XX
   d They have different lifestyles.

2 Chromosomes; information; proteins; characteristic

3 Unspecialized; embryos; research

4 **a** **i** 50% **ii** 2
   b **i**

	Sarah	
	T	t
t	Tt	tt
t	Tt	tt

(Alan)

   ii 50%

   c Any one of height, eye colour, skin colour, or many other possible characteristics

5 **a** Loss of control over movements; memory loss and mental deterioration
   b Abigail and Brenda
   c **i** His employer may not want him to work for the company for long; he may miss out on promotion.
      ii Advantage: they can consider having a termination if the test is positive. Disadvantage: deciding whether or not to have a termination is a very difficult decision.

C1 Workout

1 78% nitrogen; 21% oxygen; 1% argon; 0.04% carbon dioxide (variable water vapour)

2 Car A, going in to engine: N_2; O_2; coming out of exhaust: N_2; CO_2; H_2O
   Car B, going in to engine: N_2; O_2; coming out of exhaust: N_2; NO_2; CO_2; H_2O; NO; C; CO

3 Nitrogen dioxide – dissolves in rainwater and sea water and lowers the pH of rain; sulfur dioxide – dissolves in rainwater and sea water and lowers the pH of rain; particulate carbon – makes surfaces dirty and causes health problems if breathed in; carbon dioxide – dissolves in rainwater and sea water and used by plants in photosynthesis

4 Carbon or hydrogen; hydrogen or carbon; oxygen; carbon dioxide; water; chemical or combustion; number; products; rearranged; products

5

	Reactants		Products
Name	coal (with no sulfur impurities)	oxygen (from a plentiful supply of air)	carbon dioxide
Formula	C	O_2	CO_2
Diagram			

	Reactants		Product		
Name	coal (with no sulfur impurities)	oxygen (from a limited supply of air)	carbon dioxide	carbon monoxide	particulate carbon
Formula	C	O_2	CO_2	CO	C
Diagram					

	Reactants		Products	
Name	coal (with sulfur impurities)	oxygen (from a plentiful supply of air)	carbon dioxide	sulfur dioxide
Formula	C and S	O_2	CO_2	SO_2
Diagram				

7

Pollutant name	Pollutant formula	Where the pollutant comes from	Problems the pollutant causes	One way of reducing the amount of this pollutant added to the atmosphere
sulfur dioxide	SO_2	burning fossil fuels with sulfur impurities	acid rain	remove sulfur impurities from fuels before burning
nitrogen oxides	NO_2 NO	burning fuels in car engines	acid rain asthma	fit catalytic converters to cars
carbon dioxide	CO_2	burning fossil fuels	global warming	burn less fossil fuels
carbon monoxide	CO	burning fossil fuels in a limited supply of oxygen	poisoning	make sure oxygen supply is plentiful
particulate carbon	C	burning fossil fuels in a limited supply of oxygen	makes surfaces dirty	make sure oxygen supply is plentiful

8

A	T	M	O	S	P	H	E	R	E	W
R	N	O	X	U	E	U	C	E	S	O
G	A	N	Y	L	T	M	C	T	T	L
O	T	O	G	F	R	A	I	A	N	I
N	U	X	E	U	O	N	T	W	A	M
L	L	I	N	R	L	S	Y	D	T	P
E	L	D	N	I	A	R	L	I	C	U
U	O	E	L	E	S	S	A	C	A	R
F	P	R	O	D	U	C	T	A	E	I
D	E	G	N	A	R	R	A	E	R	T
E	L	E	C	T	R	I	C	I	T	Y

C1 GCSE-style questions

1 a A hydrocarbon is a compound made from hydrogen and carbon only.

   b i

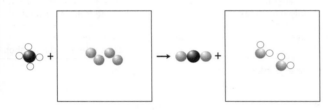

   ii It is used by plants in photosynthesis; it dissolves in rainwater and sea water; it mixes with other gases in the atmosphere.

   c i It makes surfaces dirty, or causes health problems if breathed in.

   ii Carbon dioxide = CO_2; carbon monoxide = CO; particulate carbon = C

   d Burning less fossil fuel

2 a Oxygen; monoxide; dioxide

   b Acid rain *or* increases risk of asthma attacks

3 a i

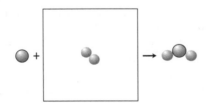

   ii Hydrocarbon fuels with sulfur impurities

   b i Increased

   ii More coal-fired power stations *or* more vehicles burning hydrocarbon fuels *or* any other sensible answer

   c i Nitrogen; oxygen

   ii Acid rain damages buildings made of limestone; acid rain makes lakes more acidic; acid rain damages trees.

   d Power stations remove sulfur from coal before burning it; British people use less electricity; power stations remove sulfur dioxide from waste gases; less electricity is being produced in power stations that burn coal.

P1 Workout

1 a C   b A   c D   d A

2 a 3 hundred thousand
   b 4 thousand million
   c 10
   d 12 700
   e 1.4 million
   f 1000

3 Words/phrases *not* to cross out are: away from us; increases; Hubble's; expanding; thousand

4 From top to bottom: C, A, B

P1 GCSE-style questions

1 a Universe, Sun, Earth

   b Moon – a natural satellite; comet – a lump of rock held together by ice; asteroid – a body that looks like a small, rocky planet; star – a ball of hot gases

   c There was less light pollution 2000 years ago.

   d i It orbits the Sun *or* it is big enough to be spherical.

   ii Its orbit overlaps Neptune's, so it has another object in its orbit.

2 a From centre: core, mantle, crust

   b 1C; 2D; 3B; 4A

3 a Milky Way

   b Thousands of millions

   c The stars of Andromeda emit light.

   d i They are at different stages of their life cycles.

   ii Everything we know about stars and galaxies comes only from the radiation they emit; Scientists use a star's relative brightness to measure its distance from Earth. But relative brightness also depends on what stage in its life cycle a star is at.

   iii The distance light travels in one year

4 a 10 cm/year

   b B because C and A *or* B because A and C

   c 1B; 2C; 3A

B2 Workout

1 Sweat, skin, tears, stomach acid

2 Top empty box: reproduce rapidly
   Empty boxes on middle line: make toxins *and* reproduce rapidly Bottom empty box: disease symptoms

3 C E F B D

4 a Experts meet in April...
   b The eggs provide food...
   c In October...
   d This flu virus is delivered...

5 a T   b T   c T   d T   e F

6 a

Part of circulation system	What does it do?	What is it made from?
heart	It pumps blood around your body.	muscle
artery	It takes blood from your heart to the rest of your body.	Arteries are tubes. They have thick walls made of muscle and elastic fibres.
vein	It brings blood back to your heart from the rest of your body.	Veins are tubes. They have thin walls made of muscle and elastic fibres.

b i If fat builds up inside the coronary arteries, a blood clot may form on the fatty lump. If the clot blocks an artery, part of the heart muscle does not get oxygen. The cells start to die and the heart is permanently damaged.

ii Any three from: cut down on fatty foods; stop smoking; lose weight if you are overweight; take regular exercise; eat less salt; if necessary, take drugs to reduce blood pressure and cholesterol level.

B2 GCSE-style questions

1 a The cough was caused by a virus.
b i They killed the bacteria that caused the painful ear.
ii Random changes (mutations) in bacteria make new varieties that are less affected by antibiotics. Some of these varieties survive a course of antibiotics. Not finishing the course increases the likelihood of bacteria becoming resistant to antibiotics.

2 a Smoking cigarettes; drinking too much alcohol; not taking regular exercise
b The heart needs a continuous supply of energy; blood brings a constant supply of glucose and oxygen to the heart.
c E D A C

3 a Stomach acid
b i Reproduction
ii 4
c i White
ii Taking anti-diarrhoea tablets would mean that *Salmonella* bacteria would stay in the intestines for longer.
d Resistant; killed; bacteria

4 a Increased: opinion B *or* C *or* D
Decreased: opinion A
b E C B D
c To protect everyone from the disease – even those who cannot be vaccinated for some reason

C2 Workout

1 Example answers are given here – many others are possible.

Part of tricycle	Properties this part of the tricycle must have	Material
tyres	high frictional force with ground	rubber
brake	high frictional force with wheel	plastic
frame	high strength	steel
seat	high strength; not cold to the touch	polypropylene
handle to push tricycle	high strength; not cold to the touch	polypropylene
pushing pole	high strength	aluminium
screws that join pushing pole to handle	resistant to corrosion	stainless steel
bag	flexible	polythene

2 a Recycle   **b** Reuse   **c** Landfill
d Using the product   **e** Cradle   **f** Use
g Grave

3

Product	Using the product	Disposing of the product
packaging – polythene bags	often use just once	can recycle, but must separate polythene from other waste
toys	use many times	easy to reuse – give to someone else or a charity shop
water bottles	often use just once	can reuse

4 Words *not* to cross out: strong; high

5 Middle row, left to right: fuels; raw materials for chemical synthesis; Bottom row, left to right: methane; petroleum jelly

6 See table on page 52.

7 a C   **b** P   **c** C   **d** P

8 a One more oxygen molecule
b Two more carbon dioxide molecules
c Three more water molecules

C2 GCSE-style questions

1 a Artificial heart valves D; hospital laundry bags C; fillings for front teeth B; contact lenses A
b i Cellulose
ii Non-toxic; flexible; high strength in tension
iii Polymerization

2 a Polypropylene is stronger under tension; does not rot.
b The environmental impact of making rope from each material; the environmental impact of disposing of rope made from each material
c Reuse – use the rope for another purpose; recycle – melt down and remould to make something else; recover energy – burn in an incinerator that generates electricity; landfill – put in a hole in the ground.

3 a i Five; water
ii In total, 3 carbon dioxide molecules
b i A compound made from carbon and hydrogen only

ii Lubricants; raw materials for chemical synthesis

4 a The forces between long hydrocarbon molecules are stronger than the forces between short hydrocarbon molecules; the stronger the forces between molecules, the more energy is needed to separate them.

b i Shower curtains, because they need to be most flexible

ii The rubber in car tyres, because it needs to be more rigid, harder, and stronger

P2 Workout

1 Microwaves, light, X-rays

2 Ionizing radiations: ultraviolet, X-rays, gamma rays
Radiations that cause a heating effect only: light, infrared, microwaves

3 The satellite both absorbs and transmits radiation. The air transmits radiation. The transmitter is a source of radiation. The TV is a detector. It absorbs radiation. The energy deposited here by a beam of radiation depends on the number of photons and the energy of one photon. The hill reflects radiation. The energy that arrives at a surface each second is the intensity of the radiation.

5 Ionizing; chemical reactions; vibrate; intensity; time

6 Carbon dioxide only: A, G, H, I
Ozone only: B, C, E, J
Both: D, F (accept I)

P2 GCSE-style questions

1 a Infrared – transmitting messages between remote controls and televisions; microwaves – transmitting messages between phone masts; radio waves – broadcasting television programmes

b Source emits radiation; glass transmits radiation; metal reflects radiation; water in food absorbs radiation.

c Metal walls prevent microwave radiation leaving the oven *or* cannot operate oven when door is open.

2 a Ultraviolet radiation is a type of ionizing radiation; ultraviolet radiation can make cells grow in an uncontrolled way; ultraviolet radiation makes molecules more likely to react chemically; ultraviolet radiation damages the DNA of cells.

b Clothes *or* sunscreen

c i Ozone molecules absorb ultraviolet radiation.

ii Ozone protects living things from the Sun's UV radiation. Anything that destroys ozone – such as some chemicals in aerosols – therefore increases the risk of humans getting skin cancer and eye cataracts.

3 a The intensity of the radiation arriving at Helen's phone is less *or* the distance between Helen and the source is greater.

b i Decreases

ii Radiation spreads over a larger and larger surface area.

c Vibrate; intensity; time

4 a i Combustion, respiration

ii Photosynthesis, dissolving in water

b i The rate at which carbon dioxide was added to the atmosphere was the same as the rate at which it was removed from the atmosphere.

ii Humans have burned more fossil fuels; humans have cut down forests and burned the wood.

c i Rising sea levels; changing climates

ii Methane, water vapour

B3 Workout

1 Natural selection only: C, E
Selective breeding only: B
Both: A, D

2 a Effector cells

b Neuron

c Central nervous system

d Neuron

e Receptor cells

5

N	T	S	Y	S	T	E	M
E	H	U	Y	E	R	S	A
U	R	R	P	X	A	P	N
R	E	V	O	L	V	E	D
O	E	I	C	O	E	C	D
N	E	V	E	R	L	I	O
S	E	A	R	T	H	E	O
S	O	L	A	R	C	S	F

B3 GCSE-style questions

1 a Electrical; brain; chemicals; slowly; long

b Nervous system changes – A, B, F; hormone system changes – C, D, E

2 a Studying fossils; analysing the DNA of modern cats, lions, and other species of the cat family

b i Selection

ii B, D, A

c i 6.7 million years ago

ii Fishing cat

iii Environmental; mutations; selection

3 a Global warming

b i Antarctic whelk *or* brittlestar *or* Trematomus bernacchii fish

ii Their populations will decrease.

c Environmental conditions change; another living thing in the brittlestar's food chain becomes extinct.

C3 Workout

1 X2D, X3F, X4G, Y1A, Y2C, Y5B, Y4E

2 Colouring – to make food look attractive
Artificial sweeteners – to make a food taste sweeter
Flavour enhancers – to make food taste better

3 Row 1 – toxic
Row 2 – toxic
Row 3 – gluten *or* peanuts
Row 4 – stored cereals
Row 5 – crisps

4 Words *not* to cross out: pancreas; insulin; adults; insulin; high; fat

5 B H I A G E L D F J C K

6 1 Fertile
   2 Potassium
   3 Consumption
   4 Decay
   5 Chemicals
   6 Harvesting
   7 Safety
   8 Standards
   9 Hormone
   10 Risk
   11 Phosphorus

C3 GCSE-style questions

1 a Flavourings – to make food taste better; emulsifiers – to mix ingredients; preservatives – to stop harmful microbes growing; antioxidants – to prevent fats and oils reacting with oxygen

   b Food additives with E numbers have passed safety tests; food additives with E numbers have been approved for use in the European Union.

2 a More than
   b i 56 g   ii 16 g
   c Small molecules dissolve in the blood more easily than big molecules; only small molecules can pass through the intestine walls.
   d Carbon, hydrogen, nitrogen, oxygen

3 a i To maximize crop yields
      ii Use natural predators to eat pests; rotate their crops regularly
   b i Taken up by plant roots; through chemical processes involving soil bacteria
      ii They dissolve in water that moves through the soil.
   c Add manure; rotate their crops; grow legumes (peas, beans, clover)

P3 Workout

1 Left-hand person: A, C, D, E; right-hand person: B, C, D, E, F

2 Ionizing; ground; space; atoms; medical treatment; ionizing; chemically; kill; damage

3 1B or C; 2A; 3D; 4F; 5C or B; 6E

5 a 65 units
   b 35%

6 a T   b F   c T   d T   e T
   f F   g F   h T   i F

7 a A and B; D and E
   b F
   c E
   d B and C

P3 GCSE-style questions

1 a i Radioactive
      ii Gamma radiation can penetrate deep inside Arthur's body to reach his tumour.
   b i Kill them
      ii The gamma radiation also kills healthy cells.
   c i To prevent ionizing radiation leaving the room
      ii To minimize the dose of ionizing radiation s/he receives

2 a The activity of the Cs-137 source decreases over time; the half-life of Cs-137 is 30 years.
   b 2.5 g
   c Unstable; stable

3 a D A C G B
   b Rate, boron, neutrons
   c Low level – pack it in drums; medium level – mix it with concrete; high level – very difficult to store safely

Ideas about science 1 Workout

1 a A   b B   c D   d A and C

2 1 data   2 range   3 outlier
   4 reliable   5 melting point
   6 inaccurate   7 mean   8 higher
   9 true   10 repeat   11 lowest
   12 solid   13 opinions   14 no
   15 silver

3 The temperature in the main part of the room is different from the temperature behind the curtain; the thermometer may be inaccurate; he may read the thermometer incorrectly.

Ideas about science 1 GCSE-style questions

1 a It is an outlier – its value is very different from the other data.
   b Range = 1.2 to 1.6 g; average = 1.4 g
   c The mean for restaurant B is outside the range for restaurant A.

2 a i 232 °C
      ii To be more confident the result is close to the true value
   b 229 to 236 °C
   c The metal was unlikely to have been tin as the value the student obtained was outside the range of the melting point measurements obtained by all the other students.

3 b i Outlier = 11
      ii Student R did not use the measuring equipment properly or student R's measuring equipment was faulty.
   c i 27 µg/m^3 (ignoring student G's result)
      ii 21 to 34 µg/m^3

4 a Wind direction varied; a nearby coal-fired power station was running on some days, but not on others; one student used the measuring instrument incorrectly; they did not use the same measuring instrument each day.
   b The percentage of damaged sperm in samples taken from people at regular intervals throughout the six months

5 a The concentrations of atmospheric CO_2 in 1990, 2000, and now
   b The concentrations of both atmospheric CO_2 and of CO_2 dissolved in sea water, in at least two different years
   c The concentration of atmospheric CO_2 might increase more slowly than the scientist predicted; shellfish might adapt and be able to make shells at a pH lower than they can now; any other sensible suggestions

Ideas about science 2 Workout

1   A, B, C, E

2   A8 larger, B3 correlation, C6 chance, D4 control, E9 mechanism, F10 random and F7 matched, G2 outcome, H1 factors, I5 causes

Ideas about science 2 GCSE-style questions

1   **a**   Factor: type of shampoo; outcome: percentage of breakage

b   Increase the number of hair samples tested; test each sample more than 3 times.

c   **i**   Person the hair is from; length of hair
ii   To make sure the test is fair

2   **a**   Factor: In 2004, Ireland banned smoking in all workplaces, including pubs. The other three statements are outcomes.

b   **i**   There is a correlation between banning smoking in workplaces and symptoms or smoke irritation/ concentration of carbon monoxide in pub workers' breath/pub workers' breathing difficulties.
ii   Some other factor may have caused the outcome.

3   **a**   The smaller the concentration of ozone in the upper atmosphere, the greater your chance of getting a cataract; there is a correlation between the concentration of ozone in the upper atmosphere and the number of people with cataracts; wearing sunglasses that protect against ultraviolet radiation may reduce your chance of getting a cataract.

b   **i**   As large as practically possible
ii   Time people spend outdoors; ethnic origin

4   **a**   Factors: A, F; outcomes: B, C, E

b   B The atmospheric concentration of sulfur dioxide 3000 m above ground may have changed very little; C The atmospheric concentration of nitrogen dioxide near the ground over southern England may have changed.

c   1C; 2D; 3A; 4B

5   **a**   **i**   Research from the 1970s...
ii   In two new studies...

b   **i**   The researchers thought that...
ii   Scientists are more likely to accept that the factor caused the outcome.

Ideas about science 3 Workout

1   Data from observations A, B, C; explanation D, E, F; prediction G; do an experiment H; study new data I; a different explanation J

2   Data from observations A, B, C, D, E; explanation F, G, H; prediction I; do an experiment J; study new data K; a different explanation L

Ideas about science 3 GCSE-style questions

1   Data: B, C, E, F; explanation: D

2   If they find iridium, this will confirm the prediction; if they do not find iridium, the prediction may still be correct (there may be iridium there that the scientists do not find).

3   **a**   A C; B D; C D; D E; E D

b   The data increase confidence in the explanation; the data agree with the prediction.

4   **a**   1 C and H; 2 A; 3 E and F; 4 D; 5 G and I; 6 B

b   It is impossible to predict when the next Earth movement will be.

5   **a**   **i**   Two from: A, B, C, E
ii   D

b   The data would not prove that the explanation is correct; the data would agree with the prediction; the data would increase confidence in the explanation.

Ideas about science 4 Workout

1   E F A D B C

2   **a** C   **b** X   **c** X   **d** C   **e** X

Ideas about science 4 GCSE-style questions

1   **a**   So that other scientists can repeat the research; so that other scientists can evaluate the research

b   **i**   Which combinations of food additives the scientists tested
ii   To check the reliability of data

c   1C; 2A; 3B

2   **a**   **i**   US scientists looked in medical journals...
ii   They found 63 reports... Each study showed...
iii   The scientists calculated...

b   The claim is probably reliable as a large number of scientists did studies and came to similar conclusions.

c   Since 1966, evidence from many scientists has shown that there is a link between vitamin D intake and cancer risks.

3   **a**   Four from A, B, D, F, I, J

b   C, E, G

c   H, K

Ideas about science 5 Workout

1   1 harm; 2 chance; 3 advance; 4 ionizing; 5 consequences; 6 ALARA; 7 precautionary; 8 actual; 9 benefit; 10 perceive; 11 acceptable

Ideas about science 5 GCSE-style questions

1   **a**   Wear goggles; use the sunbed for a maximum of 15 minutes; do not use the sunbed if you have lots of moles or freckles.

b   **i**   Having a tan makes them feel more confident; sunbeds can help to clear up acne.
ii   Actual risk B; perceived risk C

2   **a**   A I; B I; C G; D G; E I; E I

b   They like the taste; it is high in protein; it is relatively cheap compared with other meat; any other sensible answer

c   Brendan

d   For example: we will make and enforce regulations to ensure that the risk of infection by *Campylobacter* is as low as is reasonably possible. We will also ensure that the public knows what action they can reasonably take to avoid infection.

3   **a**   **i**   Paragraph 1
ii   Paragraph 2

4   **a**   Greenhouse gas emissions decreased.

b   Chances; consequences

c   Precautionary principle – the consequences of flooding could be catastrophic for these people.

Ideas about science 6 Workout

1   Ticks in boxes next to statements: B, C, F, H, K, L

2   1B; 2C; 3A

3   Words: benefits, costs, sustainable, regulations, ethical, values, questions, technically, feasible, unnatural, majority, risk

Ideas about science 6 GCSE-style questions

1   **a**   Ian, Jasmine
    b   Making sure that used fuel rods are disposed of safely; regularly checking storage sites

2   **a**   Agree: Marcus, Nikki; disagree: Kirsty, Linda, Oliver
    b   Technical issues: Does the technique damage embryos? Do embryos that have been tested grow properly? Is PGS necessary – maybe embryos can fix their own defects?
    Values: Is PGS ethically acceptable? Is it natural to choose which embryo to implant? Is it right to destroy embryos that are not implanted?
    c   Different countries have different laws related to the ethics of the process; some countries decide not to spend money on the treatment.

3   **a**   Organic farming does not use synthetic fertilizers; organic farmers control pests with natural predators.
    b   **i**   The energy in the waste is not simply released to the atmosphere as low-grade heat energy.
        ii   Costs of transporting the waste to power stations may be high; sorting waste that can be used in this way may be expensive or technically difficult.

4   **a**   Technical issues: malaria parasites become resistant very quickly; there are many different malaria parasites.
    Values: malaria mainly affects people in economically poor countries; drug companies make more money from selling drugs for heart disease than they could make from a malaria vaccine; governments of economically rich countries are unwilling to spend money on scientific research to develop a malaria vaccine.
    b   **i**   There are risks associated with the side effects of all vaccines.
        ii   Nearly the whole population would be protected from mumps, measles, and rubella if the MMR vaccine was compulsory.

Index